# The Politics and Technology of
# Satellite Communications

# The Politics and Technology of Satellite Communications

Jonathan F. Galloway
Lake Forest College

**Lexington Books**
D.C. Heath and Company
Lexington, Massachusetts
Toronto          London

**Library of Congress Cataloging in Publication Data**

Galloway, Jonathan F
The politics and technology of satellite communications.

    Bibliography: p.
    1. Artificial satellites in telecommunication.
I. Title.
HE9721.U5G35        384.54'56       72-5238
ISBN 0-669-84467-5

Published simultaneously in Canada.

Printed in the United States of America.

International Standard Book Number: 0-669-84467-5

Library of Congress Catalog Card Number: 72-5238

To the memory of my father,
George Barnes Galloway, 1898-1967.

# Contents

List of Tables and Figures      xi

Acknowledgments      xiii

Chapter 1      **Introduction**      1

Conceptions of American Policy      2
Clarification of Purposes      4

Chapter 2      **Traditional Communications
Policy and the Coming of the Space Age**      9

Participants in the Policy Process      11
The National Aeronautics and Space Act
   and the Traditional Communications Structure      12
Conclusions      15

Chapter 3      **Early Developments in Space Communications**      17

Frequency Allocations      17
Civil-Military Relations      19
The Adequacy of Policy and Policy-Making Machinery      21
Government-Industry Relations, 1960-1961      22
Attitudes of Business      27
Attitudes in the Executive Branch      29
Congressional Attitudes      32
Technical Characteristics of Alternative
   Communications Satellite Systems      35
The Question of Technological Breakthrough      37
Demands for Communications Satellite Service      39
Characterization of One Stage of the
   Decision-making Process      41
Conclusions      43

Chapter 4      **The Passage of the Communications Satellite
Act of 1962**      47

Drafting of the Legislative Proposal Within the
    Executive Branch      47
Congressional Attitudes, Hearings
    February to April, 1962      52
Attitudes of the Executive During the Congressional
    Hearings, February to April, 1962      57
The Attitudes of Interest Groups      62
Committee Reports and Floor Action      64
Conclusion      69

Chapter 5      **The Interim Arrangements for a Global Commercial
Communications Satellite System**      75

Frequency Allocations and Regulations for
    Space Communications      75
The Establishment of INTELSAT      80
Conclusion      99

Chapter 6      **Space Communications and National Security**      105

Projects West Ford and ADVENT      105
Defense Negotiations with Comsat, 1963-1964      106
The Defense Satellite Communications System      113
Foreign Relations Aspects of National Security
    Requirements for Communications, 1965-1972      114
Conclusions      118

Chapter 7      **INTELSAT and INTERSPUTNIK: American-Soviet
Relations Concerning Space Communication**      121

Soviet Attitudes and Plans for Satellite Communications      122
Cooperation      124
Competition      127
Conflict      132
Conclusions      134

Chapter 8      **The Transition Between the Interim and Definitive
Arrangements for INTELSAT**      137

Domestic Politics of Foreign Policy Making 137
Assessment of INTELSAT, 1964-1972 147
Negotiation of the Definitive Arrangements 155
Distribution and Direct Broadcast Satellites 164
Conclusions 168

**Chapter 9**    **Conclusions** 171

Policy Goals and Policy-Making Processes 171
"Muddling Through" 174
The Impact of Technological Innovation 176
The Boundaries of Domestic and Foreign Policy 180

**Appendixes** 185

Appendix A 187
Appendix B 193

**Notes** 199

**Glossary** 227

**Bibliography** 231

**Index** 243

**About the Author** 249

# List of Tables and Figures

**Table**

3-1          Projected Needs for Voice Channels                                    40

8-1          Government Estimates for Communications Satellite
             Research and Development Program                                      144

8-2          Structural-Functional Relationships in
             INTELSAT                                                              151


**Figure**

8-1          INTELSAT IV and Earth Station Locations
             in 1972                                                               152

# Acknowledgments

My father and mother, George and Eilene Galloway, have provided constant stimulus and encouragement to my scholarly pursuits, not only in this book, but throughout my life. Without the pervasive atmosphere of interest in understanding and reforming politics in our Washington home, my own knowledge of politics, in particular Congressional politics, would be much poorer. My wife, Judy, has been footnote assistant, a thankless task. More than that, she has been my compassionate spiritual companion. Our three joyful and rambunctious children, Matthew, Jennifer and Anne, were born during the time I was working on this book, which I first began researching as a doctoral dissertation in 1964. While much of my debt for information goes to voluminous Congressional hearings, I could not have explained the politics and technology of satellite communications without numerous interviews. In particular, William Gilbert Carter, Stephen E. Doyle, and Edwin J. Istvan were most perceptive and encouraging. However, it goes without saying that the analysis and interpretation which follow are my own.

# The Politics and Technology of
# Satellite Communications

# 1  Introduction

Of all the technologies modern civilization has produced, the one which has most saturated industrial societies, and transformed traditional societies, has been communications technology. Changes in communications techniques, from the printing press to television, have been associated with the birth of democracy, mass societies characterized by anomie and alienation, and total ideological war. The analyst interested in an inductive approach, rather than intuitive generalizations, must understand the discrete processes through which individual technologies in communications, transportation, production, medicine, or weapons have been introduced into national and international life. He must comprehend the consequences of each innovation, as well as the combined and interrelated effects of all innovations, so that he may understand the modern world and the choices man has for improving his life.

The present study approaches this general task by examining one technological innovation—satellite communications—as it relates to three basic processes:

1. The relation of innovations in communications technology to innovations in policy and the policy-making process
2. The distinction or lack of distinction between domestic and foreign policy decision-making
3. The processes of rational decision-making characteristic of and appropriate for issues which are complex because of the consequences of technological innovations

These processes, which are the three themes of this book, cannot be understood in isolation. It will be demonstrated that, in the case of space communications, they are intertwined largely as a result of the impact of technological change and as a result of man's attitudes toward controlling that change.

We will analyze these three themes through a historical treatment, and through contemporary interpretation of the issues, participants, and major events which have made the politics of space communications an important part of existing politics within and between nations. Basic data necessary to our task will be provided.

Though the innovative process is a more or less continuous one, for the purposes of analysis we identify stages in the process and examine specific innovations (cable, point-to-point communications satellites, direct broadcast global television, etc.). We believe this approach legitimate and necessary.

1

The impact of the mass media on public opinion, and attitudes toward it by elites and officials—aspects of our first theme—have often been the subjects of study by social psychologists, sociologists, and political scientists;[1] but when it comes to analyses of particular innovations and their place within the political life of a country, the studies are fewer.[2] Furthermore, while there are many studies on changes in the means of production, the means of distribution, and weapons technology, there are relatively fewer about the telephone, telegraph, cables, radio, and television. The printing press has received its share of academic attention as a contributor to nationalism and individualism, but no one would deny that more study should be devoted to the effects of innovation in communications technology. For political scientists, an analysis of the diplomatic, military, economic, and psychological instruments of public policy would be seriously incomplete without an examination of the modern means of communications.

Our second theme, interaction between the making of domestic and foreign policy, is closely related to the first. Decision-makers may have different roles, functions, and sources of information in the two arenas; their constituencies in the implementation of policy are often pluralistic, having overlapping or coincidental interests rather than common interests. Sorting out the threads and strands of policy-making offers a complicated task to the student of political behavior, especially since the analyst may very well complicate the process by becoming a part of it. But to the degree that objectivity is a realizable goal, it will be sought, for it offers a clear view for understanding developing alternatives and their probable consequences.

In approaching the third theme—the process of rationality in decision-making—it is important to consider what is meant by the term "rational" and how participants evaluate their behavior in an environment which is constantly changing. We will analyze how criteria of rationality appropriate to a stable environment and limited problems, may not be appropriate to a changing environment with novel and ambiguous problems.

## Conceptions of American Policy

Space communications policy has developed as part of overall American policy devoted to certain general goals, foreign and domestic. Components of the American view of life and the future are of course subject to contrary interpretations in the context of particular situations—this is the meat of politics—but it is well to list at the beginning those general conceptions of policy about which there is a considerable degree of consensus. These general policies will be interpreted in the concluding chapter, in view of the concrete happenings in space communications between 1957 and 1972.

American policy is verbally committed to:

1. Encouraging peace and understanding among nations
2. Increasing national security
3. Increasing trade and commerce between nations
4. Assuring American leadership in science and technology
5. Strengthening international organization
6. Working for arms control and disarmament
7. Aiding underdeveloped countries
8. Projecting the free enterprise system
9. Preserving and extending the democratic way of life at home and abroad.

These aims are the political parameters within which all specific policies have been shaped. United States satellite communications policy must be seen not in isolation, therefore, but as a small part of overall American policy, related to the broad objectives of our foreign, traditional communications, and space policies. With this in mind, let us elaborate on some of the many issues which touch on space communications developments.

One highly sensitive domestic issue is that of the relative roles of government and industry in the making of policy. Who are the major participants within each sphere and how do their attitudes interact to influence policy? This is an especially crucial question in space communications, which involves foreign relations. How are the various participants' preferred alternatives made attractive to different types of clients, domestic and international? In what ways are currently-popular concepts of analysis relevant to an understanding of the place of the Communications Satellite Corporation (Comsat) in the American scheme of things? Does treating Comsat as a private corporation clarify the situation or obscure it? Does calling Comsat the chosen instrument of American foreign policy clarify or obscure? And, *how* do they clarify? *How* do they obscure?

Another important problem is the role of the Department of State in the conduct of American foreign relations. While many grant the Department a central role, the problems of interdepartmental coordination and competition destroy whatever hopes exist for a hierarchical organization of responsibilities. The Constitution, as Corwin writes, is "an invitation to struggle for the privilege of directing American foreign policy;[3] but, even beyond this, the informal and shifting bargaining coalitions can be an invitation to formidable complexity and confusion. In addition, the use of the Department as a whipping boy by certain elements in Congress and the public, plays havoc with any image of State as a powerful force in the internal decision-making process. It is thus relevant to ask what the role of the Department was in the formulation and implementation of American policy for space communications. What conditions have made for an active role for State, and which conditions for a less pronounced place in the policy process? How cohesive has the Department itself been in various situations?

Space communications has involved the United States in relations with industrialized allies and neutrals, developing countries, and Communist states. Hence, space communications policies are relevant to an understanding of certain major trends in relationships between these types of states.

For instance, one perennial concern is the character of U.S. relations with the West European allies. The relationship of Comsat and the Department of State and other governmental departments to a definition of policy in this sensitive area, involving perceptions of American predominance and partnership, will shed light on a small part of the problems of transnational cooperation and integration. The effects of nationalism and historical factors which have conditioned world communications traffic will be assessed in this connection. What has been the attitude of Britain towards the introduction of a new technology whose potential could dwarf the British stake in cables? How has France's search for glory and prestige influenced international arrangements for space communications? To what extent has Europe become an integrated actor in this area?

The relationship of developing countries and industrialized nations will be described and assessed. Can early planning and initiative in space communications provide an important piece of infrastructure for the newer nations? If so, how has American policy reflected this possibility? What are the needs of the less developed countries for space communications? How do representative government officials estimate the requirements of the new states? What has been accomplished so far?

Developments in space communications are also involved in East-West relationships. An understanding of these developments will provide partial data about the nature of the change in the Cold War. To what degree has the Soviet Union cooperated with the United States in international space communications endeavors? Is the bipolarity which is characteristic of the contemporary strategic military system characteristic of the international communications system? How concrete has American policy been in seeking the goal of increased cooperation and understanding? Is ambiguity in goals a result of the disturbing consequences of technological innovation or a result of perennial factors?

## Clarification of Purposes

Interpretation along lines of the three themes stated above will first be presented in connection with a brief historical survey of United States policy as related to communications innovations in the telegraph, undersea cable, and radio. Following this historical introduction, we will analyze in rough chronological order the course of events, issues, and participants surrounding developments in space communications. Before turning to the main body of the book, however, let us elaborate how a consideration of these three tasks will contribute to our general knowledge of political behavior.

The main concentration will be on analysis of the first theme, that is, the nature of a technological change in communications and its relation to innovations in policy and the policy process. Discussion of the pace and character of technological change in general has been widespread. In John von Neumann's off-quoted words, "all experience shows that even smaller technological changes than those now in the cards profoundly transform political and social relationships."[4] Following from this assessment comes the prescription that we must plan our future with more care than we have done in the past. But the requirements of planning may interfere with the traditional philosophy of private enterprise: they may upset the stakes of those with vested interests in exploiting new technologies. On the other hand, the consequences of unplanned technological change may disturb tradition more profoundly and chaotically.

One line of inquiry is to ask how advances in communications differ from advances in military, production, transportation, and health technologies. This is a difficult question to isolate, for two reasons: (1) differences presumably could follow disparate patterns in different cultures and at different times; and (2) advances in one type of technology are related to advances in others. In fact, the above names for "technologies" are only analytic devices, useful for some purposes, misleading for others. Production can be improved from a technical standpoint by both automation and cybernation. The railroad is a means of transportation, communication, and a tool for war or defense.

As this study is not a comparative analysis of different technological innovations, no attempt will be made to provide a comprehensive picture of satellite communications technology as compared to other technologies. But information related to this picture will be given as it relates to the expectations of the policy makers responsible for American policy for satellite communications. That is, we will list the goals and expectations of the important decision makers and then relate them to actual developments. One by-product of this procedure will be to indicate the differences and/or similarities between the expected uses of space communications systems and their social effects.[5] In this way we will be able to differentiate between the uses of space communications per se and their impact on production, defense, education, and mass attitudes towards international peace and understanding.

In order to place the changes in satellite communications technology within the context of changes in technology in general, it will also be helpful to distinguish three general images decision makers and other observers have on the subject of technological change in the contemporary world. We can then test these images against the actual developments in space communications.

Image one treats technological change and its increasing rate as a profound and disturbing natural force. Image two, on the other hand, treats it as a social invention—seeing no technological change without changes in the political and social context. The third image falls at a point between the other two, viewing technological change both as a natural force and a social invention.

To see which image more adequately fits the reality, we shall study the significance of a set of changes in communications techniques. Successive interpretations of participants, issues, processes, and environments will be undertaken, so that we may appreciate any increase in awareness of the changes associated with, and perhaps caused by, technological change in satellite communications. Concomitantly, it will be possible to evaluate the degree of dislocation which has characterized the introduction of space communications systems into the traditional communications environment. We will then possess a description of how profound the change has been.

Our second theme relates to the interaction of processes, personalities, and substantive issues between the foreign and domestic environments. The meaning and extent of overlapping and common problems varies: war, for instance, is primarily an international concern, education a domestic one, while internal revolution may have more effect on world political alignments than a war—compare the Russian Revolution to the Mexican War. It is obvious that the distinctions between domestic policy, foreign policy, and international politics are matters of analytical convenience which explain certain facets of reality at the expense of obscuring others.

The contribution to knowledge which the present analysis will provide is not in making the point that certain foreign policies affect domestic politics and vice versa. This is well known. What needs to be emphasized is not that foreign participants indirectly influence the selection of alternatives according to the "rule of anticipated reaction,"[6] but that foreign participants can *directly* take part in the policy-making process as well as the policy-implementation process. This can happen when it is believed that the effects of decisions will be domestic, foreign, international, and transnational.

It is traditionally said that foreign policies must be made in a secret, centralized, and speedy fashion.[7] More recent analysis has shown, however, that certain foreign policies have been the result of legislative politics involving bargaining between different, dispersed groups, rather than the result of executive politics involving hierarchic decision-making. It has also been shown that different types of foreign policies and domestic policies are characterized by different decision-making patterns.[8]

One can go even beyond this point. The distinction between foreign policy-making and domestic policy-making, should be qualified to allow for different types of decision-making patterns in policies which are not primarily foreign or domestic but a combination of both. This third arena of politics has been called the "penetrated system."[9] The idea that the territorial distinction between foreign and domestic policy has become less helpful as a guide to understanding politics is related to the relative decline of territory as an important variable in world politics,[10] and the increasing recognition of functional participants which transcend national political boundaries and regional organizations, which are in some cases supranational in character.

These changes in the international system mean that the content and the rules of the game have been changed, and thus have patterns of rationality—the substance of our third theme. If the rules of bureaucracies or the perceptions of policy-makers are constrained by traditional distinctions separating domestic and foreign politics, and do not reflect awareness of the penetrated system, then, to the degree that this is so, one sees less rational decision-making than is possible. Rationality, then, is a matter of degree. Further, where there is a lack of awareness of the complexities and nuances of reality—even if the goals are clearly and consistently stated—there is going to be confusion and lack of understanding. This context of policy-making is certainly affected by the tendency of technological innovations to complicate the politician's balance sheet. As Karl Deutsch has said, technological "developments have changed the number of participants in the political games, the stakes of politics, and in some basic regards the nature of politics itself.[11]

Two contending approaches to rational decision-making have been offered to deal with the increased complexity of modern society (which is caused to a great degree by accelerating and ubiquitous technological changes). One is the traditional American pragmatic approach or, as it is called in political science, incrementalism, or "muddling through." It is the rationality of a pluralistic political system such as that described by James Madison, David Truman, Robert Dahl, and Charles Lindblom.[12] Partisans of this approach contend that increased complexity justifies a go-slow approach to problem solving. When the gap between the ends desired and the effects of technological innovation in the real world can only be roughly estimated, to "muddle through," to proceed incrementally and pragmatically, may be the better part of wisdom. Otherwise, comprehensive, revolutionary decisions will be made on the basis of inadequate information and lack of normative criteria for making such decisions.[13]

Opponents of incrementalism say that it is a status quo rationality, which may at one extreme lead to catastrophe. For instance, Paul Ehrlich foresees the extinction of the human species through a convergence of the unintended consequences of all technological changes in an avalanche of time lag, synergistic, threshold, and trigger effects.[14] Herbert Marcuse maintains that the tenets of American society lead to a technological rationality which alienates man from himself and other men.[15] And at the level of bureaucratic decision-making rather than the imperatives of the general culture, political scientists have often called into question the ad hoc improvisations characteristic of American pragmatism.[16] These commentators say that we are living on borrowed time, that, we cannot afford to tinker with the machinery but must overhaul our national priorities and our decision-making procedures. Partisans of this more comprehensive type of decision-making, central planning, and systems analysis may view the present system as dominated by a power elite or a ruling class rather than a pluralistic universe of more or less equal and overlapping interest groups; although this is not necessarily the case, because proponents of a new rationality for American politics can also see traditional Madisonian pluralism as an anachronistic system of "interest group liberalism."[17]

This work contends that a solution to this competition between the incremental and comprehensive approach to rationality, insofar as technological change is concerned, has as its basis an examination of the revolutionary or evolutionary character of discrete technological innovations and their possible combinations. Certain technological changes are compatible with a "muddling through" approach to policy-making while other, more revolutionary changes require comprehensive central planning.

With this introduction, let us now proceed to a detailed examination of the politics and economics of satellite communications technology.

# 2

## Traditional Communications Policy and the Coming of the Space Age

We need to analyze three aspects of traditional communications policy which led to the National Aeronautics and Space Act of 1958. The first of these is the relationship between government and industry. In most foreign countries the government owns and operates the telegraph, telephone, and broadcast media. In the United States, a pattern developed whereby industry was subsidized by the government, in its ownership of the various means of communication. Invented during the 1830's, the first operational telegraph in the United States was constructed by the Post Office Department; but in 1857, Congress was unwilling to continue the financing of this line and turned control over to private business.[1] By the Post Roads Act of 1866,[2] Congress further assisted private development of the telegraph by making available public lands for lines. International telegraph communications through underseas cables were also owned by private interests and assisted by the government. President Grant in 1875 made it American policy to discourage foreign telegraph interests from laying cables to the United States unless these countries permitted American companies to do the same through their territory.[3] What we see in the nineteenth century is a practice developing of private ownership of the means of communication, both domestic and foreign.

At the beginning of the twentieth century, a second relationship made itself felt in conjunction with America's rise to great power status—and this was the interaction of commercial and military concerns. When Great Britain advocated the laying of an "All Red" cable across the Pacific in 1900, and then laid it in 1902, both Presidents McKinley and Roosevelt responded by encouraging American companies to lay a competing cable. In fact, if private enterprise had been unwilling to meet this challenge, both Presidents would have backed a government-owned venture.[4]

During World War I, Congress authorized the President "to supervise, or take possession and assume control of any telegraph, telephone, marine cable, or radio system . . . for the duration of the war."[5] Even before American entry into the war in 1917, the government was assisting private companies in the research necessary to improve communcations capabilities. Out of this wartime centralization of effort came technological innovations in radio which the government wanted to keep secret for national security purposes. After the war, the Secretary of the Navy proposed government ownership of radio technology, but Congressional opposition necessitated return of the radio stations to their owners in 1920.[6] However, the Navy brought pressure to bear on private

9

interests, not to disseminate abroad new technical products such as the Alexanderson alternator. Private companies were persuaded to combine wartime innovations in radio in the formation of a new company, the Radio Corporation of America, on whose board the Navy had a representative throughout the 1920's.[7]

During World War II, the means of international communcation were again put under government control. As in World War I, this greater centralization of management resulted in increased technical innovations and communications capacity. The end of the war revived the issue of whether there should be government ownership, or, at least, merger of all the means of international communication, so as to serve as a more effective instrument of both public and private interests abroad. Hearings before the Senate Commerce Committee indicated a disagreement between partisans of increased government control for national security purposes and advocates of the traditional free enterprise position.[8] The established philosophy predominated. The government-operated system was cut back drastically in the immediate postwar thrust towards demobilization.

The onset of the Cold War once again witnessed increased government involvement with communcations capabilities. Not only did the military services build up their own in-house capabilities, but the military used a major portion of private communications channels. Also, communications were coming to be used more and more for propaganda purposes.[9] But coordination of the private and public sectors and coordination within the government were major problems. The Presidential panels and Congressional committees which considered these problems during the administrations of Presidents Eisenhower and Truman, were generally dissatisfied with the effectiveness of United States' communcations capabilities.[10]

The increase in American commitments around the globe and the domestic growth of companies motivated to explore the various communications techniques, brought about the third relationship dealing with traditional communications policy. This relationship was the regulation by the government, of private common carriers not only in time of war and Cold War, but to promote efficient and fair service to the general public in time of peace. While the initial government relation to the carriers was one of assistance to fledgling companies, the growth of these companies to maturity was accompanied by regulation both domestically and internationally.

Domestically, the power to control rates was centered in the Interstate Commerce Commission by the Mann-Elkins Act of 1910, but this power lay dormant. In 1927, the Federal Radio Commission was created to control rates and frequencies for radio communcation, but it was not until the formation of the Federal Communications Commission (FCC) in 1934 that regulatory authority over all civilian means of communcation was centralized and extended.[11] The scope of this regulatory authority includes frequency allocations, rates, and establishing of communications services.

Internationally, regulation of American public and private communications has been undertaken on a technical level by the International Telecommunications Union (ITU). In the nineteenth century, the United States was unwilling to accept membership in this specialized agency, then called the International Telegraph Union, because American companies did not wish to be bound by rules which they thought would needlessly increase the costs of communcations. However, in 1912, the government joined the International Radio-Telegraph Union whose main function was to regulate ship-to-shore service for emergency purposes. At the Madrid Conference of 1932, the Telegraph and Radio-Telegraph Unions were combined into the ITU, but private interests remained opposed to international regulation of telegraph and telephone communications; consequently, the government restricted its ITU membership to the radio convention.[12] In 1949, however, the United States signed the telegraph regulations. International regulation is restricted to technical matters which must be coordinated if there are to be any communications at all. Domestic regulation is stronger because it involves rate regulation as well.

## Participants in the Policy Process

Having briefly examined the three relationships which form the basis of traditional communications policies, it is now appropriate to designate the participants in the policy-making process. A broad designation would include the Executive branch of government, Congress, the business community, international organizations, and foreign nations. Within the Executive branch, the principal departments and agencies responsible for space communications policy, in addition to the President, are the National Aeronautics and Space Agency (NASA), the National Aeronautics and Space Council (NASC), the Office of Telecommunications Policy (OTP), the Defense Department, the Department of Justice, and the Department of State. The Federal Communications Commission, to some degree an independent regulatory commission, has an important part in the policy process. Congressional committees dealing in communications are as follows: in the Senate—the Committee of Aeronautical and Space Sciences, the Commerce Committee, the Select Committee on Small Business, and the Foreign Relations Committee; in the House—the Committee on Science and Astronautics, the Interstate and Foreign Commerce Committee, and the Committee on Government Operations.

Within the business community, communications interests can be classified into those of the international common carriers, domestic common carriers, television and radio broadcasters, and manufacturers. The Bell System encompasses three subtypes, the American Telephone and Telegraph Company (AT&T) being an international and domestic common carrier, while the Western Electric Company is the manufacturing entity. Other important international common carriers are Western Union International (WUI), International Telephone and

Telegraph Company, World Communications (ITT World Com), Radio Corporation of America, Global Communications (RCAGC), Hawaiian Telephone Company (HTC), and Press Wireless. The domestic common carriers are many, but here the two big giants, AT&T for telephone and Western Union (WU) for telegraph, may be mentioned. Television and radio companies include the American Broadcasting Company (ABC), the National Broadcasting Company (NBC), and the Columbia Broadcasting System (CBS). Manufacturers include RCA, Philco, General Telephone and Electric Company (GT&E), Lockheed Aircraft Corporation, Hughes Aircraft Company, and the General Electric Company (GE).

The international organizations which have been concerned with space communications are the United Nations (UN) and certain specialized agencies, particularly the ITU. Foreign nations have been involved with American space communications policy, but did not play an active part in the policy-making process leading up to the passage of the Communications Satellite Act in 1962. They have, of course, participated in the activities of the UN and the ITU.

## The National Aeronautics and Space Act and the Traditional Communications Structure

Space communications policies reflect a converging of two overall types of policy, i.e., communications and space. Having briefly considered traditional policy and its participants, it is now appropriate to examine the circumstances and initial policy formulations of the space age. These, of course, take in a much broader scope of activities and technologies than communications per se.

On October 4, 1957, the Soviets launched Sputnik I. The United States suffered a serious blow to its image as the scientific and technological leader of the world. The impact was registered by public opinion in allied countries as well as within the United States. The challenge affected the public's image of our military superiority and the ability of the Republican Party to conduct foreign policy.[13] The irony of the situation is that the United States could have orbited a satellite in 1956 if it had made a go-ahead decision.[14] At that time, however, the policy goal of our space program was to project a peaceful image of the United States.[15] Thus, satellite development was given priority in connection with the International Geophysical Year. Responsibility for launching was assigned to a Navy missile, the Viking, which had no conceivable military uses; the Army's missile, the Jupiter, was rejected because its development was linked to the development of the deterrent forces of the United States. With the shock of Sputnik I, however, the Jupiter became a space launcher. The first American satellite, Explorer I, was orbited on January 31, 1958, by a Jupiter.

The Soviet challenge in space led our government to revaluate America's space program. In the Senate, Senator Lyndon B. Johnson, (D., Tex.) chairman of the Preparedness Investigating Subcommittee of the Armed Services Committee, started extensive hearings on November 25, 1957, into the missile and satellite program.[16] In addition, the Senate established the Special Committee on Space and Astronautics on February 6, 1958. In the House, the Select Committee on Astronautics and Space Exploration was created on March 5. The membership of these last two committees was composed of the ranking majority and minority representatives of the most concerned standing committees. The Senate Committee was chaired by the Majority Leader, Senator Johnson, and the House Committee by Majority Leader, Hon. John W. McCormack (D., Mass.).

The Executive branch was equally responsive in its reaction to Sputnik. On November 7, 1957, President Eisenhower created the post of Special Assistant to the President for Science and Technology. He transferred the Science Advisory Committee from the Office of Defense Mobilization to the White House, with the new title of the President's Science Advisory Committee. In the Defense Department, the Secretary announced the creation of a separate new agency responsible for research in the satellite and space fields. This came to be known as the Advanced Research Projects Agency (ARPA).

Within the Congress, hearings of the Special and Select Committees did not begin until the Administration submitted a proposal on April 14, 1958,[17] providing for the creation of a National Aeronautics and Space Administration to be built upon the basis of the National Advisory Committee for Aeronautics (NACA). The new agency would considerably expand on the experimental activities of NACA. What its relation to the Department of Defense would be, was a matter of much discussion in the hearings. Congress expressed concern about the respective jurisdictions of the military and the new civilian agency. It considered the Executive proposals as an imprecise attempt to meet these problems[18] and proposed modifications of the original bill. The use of space was not discussed in any depth during the hearings. One can only gain a vantage point on this issue by relating the Declaration of Policy and Purpose in the National Aeronautics and Space Act to traditional communications practice and the potentiality of space communications. The Declaration is as follows:[19]

Sec. 102 (a) The Congress hereby declares that it is the policy of the United States that activities in space should be devoted to peaceful purposes for the benefit of all mankind.

(b) The Congress declares that the general welfare and security of the United States require that adequate provision be made for aeronautical and space activities. The Congress further declares that such activities shall be the responsibility of, and shall be directed by, a civilian agency exercising control over aeronautical and space activities sponsored by the United States, except that activities peculiar to or primarily associated with the development of weapons systems, military operations, or the defense of the United States (including the research and development necessary to make effective provision for the defense of the United States) shall be the responsibility of, and shall be directed by, the Department of Defense; and that determination as to which such agency has responsibility for and direction of any such activity shall be made by the President in conformity with section 201 (e).

(c) The aeronautical and space activities of the United States shall be conducted so as to contribute materially to one or more of the following objectives:

(1) The expansion of human knowledge of phenomena in the atmosphere and space;

(2) The improvement of the usefulness, performance, speed, safety, and efficiency of aeronautical and space vehicles;

(3) The development and operation of vehicles capable of carrying instruments, equipment, supplies, and living organisms through space;

(4) The establishment of long-run studies of the potential benefits to be gained from, the opportunities for, and the problems involved in the utilization of aeronautical and space activities for peaceful and scientific purposes;

(5) The preservation of the role of the United States as a leader in aeronautical and space science and technology and in the application thereof to the conduct of peaceful activities within and outside the atmosphere;

(6) The making available to agencies directly concerned with national defense of discoveries that have military value or significance, and the furnishing by such agencies, to the civilian agency established to direct and control nonmilitary aeronautical and space activities, of information as to discoveries which have value or significance to that agency;

(7) Cooperation by the United States with outer nations and groups of nations in work done pursuant to this Act and in the peaceful application of the results thereof; and

(8) The most effective utilization of the scientific and engineering resources of the United States, with close cooperation among all interested agencies of the United States in order to avoid unnecessary duplication of effort, facilities, and equipment.

(d) It is the purpose of this Act to carry out and effectuate the policies declared in subsections (a), (b), and (c).

The Act seems to divide responsibility for aeronautical and space activities between NASA and the Defense Department. This decision, if applied to space communications, would indicate a revolutionary change from past American policy which legitimizes the role of private enterprise in the bulk of American communications. The role of the Government is regulatory, through the Federal Communications Commission. If the Act restricts civilian space activities, including communications, to the control of NASA, it may be inconsistent with the Federal Communications Act of 1934, which obligates the FCC to promote private enterprise.[20]

A different interpretation of the Act, however, makes control over space activities the prerogative of others besides NASA and the Defense Department. In the Act, "aeronautical and space activities" are defined as:

(A) *research* into, and the solution of, problems of flight within and outside the earth's atmosphere,

(B) the development, construction, testing, and operation for *research purposes* of aeronautical and space vehicles, and
(C) such other activities as may be required for the exploration of space.[21]

The dominant thrust of this definition is that, within the context of the Act, aeronautical and space activities are research activities, not operational activities. Thus, the roles of NASA and the Defense Department in controlling these activities do not prejudice the role of common carriers in operational communications activities. In fact, a pragmatic approach to decision-making would imply that private enterprise would retain its traditional place in the exploration of the new technology. The roles of the Government would be (1) to promote the development of technology for the carriers to use, and (2) to regulate an operational private system once it was established.

One may conclude, from this legal analysis, that the challenge of the Soviet Sputnik produced no threat to the existing relationship between Government and business in providing communications services to the public.


## Conclusions

The role of the Government changed from helping fledgling telegraph companies to regulating giant monopolies such as AT&T and Western Union. Military responsibilities changed from wars against Indians to the interests of the United States as a global power. On the political level, communications innovations made possible the establishment of a world-wide propaganda apparatus, as well as furthering traditional functions such as diplomatic communications and commercial relationships. In addition, the United States enunciated the policy of using the communications media as instruments in the service of world peace and understanding.

The public communications systems are not controlled or operated by the Government, as is the case in most other countries; although there are provisions for such control in times of national emergency. Hence, we have a great number of common carriers with a few giants among them, with some exercise of regulatory authority by the FCC, in order to achieve harmony in the "public interest." This system has had a profound effect on the United States in its foreign relations.

The rationale of American policy is that it promotes competition which results in lower costs to the consumer and increased technological innovations. As evidence, partisans of this theory often point to the fact that the American communications system is the best in the world. Whether this is the result of competitive private enterprise, government regulation without ownership, or some other variable, requires further analysis. American leadership in international telephone services is associated with AT&T, which holds not only a monopoly of overseas telephone services, but, has a domestic monopoly as well. Only in overseas telegraph, data-transmission, and other record services does a competitive system prevail.

American leadership in international communications was challenged by the Soviets when they orbited Sputnik I. Into the communications environment, a new technique presented itself with potentially revolutionary consequences for communications. How America responded to the Soviet challenge and the technological innovation is the subject of the following chapters.

# 3 Early Developments in Space Communications

Early expectations regarding the uses and effects of developing technology in space communications provided a seedbed which, combined with the enertia of established patterns, produced our present policies. In the previous chapter, a framework for understanding early developments was presented in terms of highlighting the basic relationships of traditional policy, the specific participants in the policy-making process, and early expectations on space policy insofar as they were revealed in the National Aeronautics and Space Act. In this chapter we will examine the complex issues which arose because of developments in communications satellite technology and their relation to the established patterns of behavior. The issues we will examine are as follows:

1. The frequency spectrum as a scarce national and world resource which needs to be carefully managed and allocated between the expanding communications services
2. Civil-Military relations as exhibited by the roles of NASA and the Department of Defense
3. The adequacy of the policy and the policy-making machinery in space communications
4. Relations between Government and industry in the development and operation of a space communications system
5. The technological systems or system best suited to bring rapid development of a global system
6. The question of whether there was a technological "breakthrough"
7. The question of the future demand for communications services via satellite

## Frequency Allocations

The first actions of the Government relating to space communications came in the field of radio frequency allocations. The Atlantic City Radio Regulations of 1947 had made no mention of space services. Thus, until 1959, all experimentation in space had to be conducted so as to cause no harmful interference to services operating in accordance with the table of frequency allocations. This was a serious handicap to activities during the International Geophysical Year and later experimentation by the United States.[1] This gap in authority led the

17

United Nations, through its Ad Hoc Committee on the Peaceful Uses of Outer Space, to call for allocations for space services by the ITU at its 1959 Radio Administrative Conference. The July, 1959 Report of the Committee urged that spectrum assignments be made for tracking, telemetry, and research purposes to cover a period of three years.[2]

In the United States, preparations for this conference had begun as far back as 1955 within the Interdepartmental Radio Advisory Committee (IRAC).[3] In 1957, Docket 11997 of the FCC was concerned with frequency allocations, some of which would be for space.[4] And, in Docket 12263, information was presented by industry giving its attitudes on the revision of the Radio Regulations of the ITU.[5] The Department of State had the responsibility, through its Telecommunications Coordinating Committee, of gathering these various inquiries in time for the conference. It arranged for consultations between United States experts and their counterparts in other countries.[6]

A possible domestic handicap to allocations for space services may have come in 1956 when the FCC began proceedings in connection with Docket 11866. The issue concerned the advisability of licensing private microwave systems which would operate separately from the established carriers. AT&T and WU objected to the possibility for many reasons, but the one that concerns us here relates to space communications. The contention was that the allocation of frequencies to a new system would take away frequencies which would soon be needed for space communications.[a] After lengthy proceedings, the FCC decided to license and allocate frequencies to further the growth and development of conventional, domestic microwave systems. However, the Commission stated that "if future conditions warrant a reallocation of frequencies to provide for space communications needs, we will then take whatever action will be necessary."[7]

Commissioner T.A.M. Craven issued a partial dissent to the opinion by declaring that it lacked a long-range perspective: "Should the Commission by sheer weight of reason, as well as public and Congressional opinion, be forced within the next few years, as I think it will, to grant space communication a preferred position in the portion of the spectrum in question here, that will mean the dispossessing of entrenched business enterprises before they have had time to amortize their investments."[8] Craven went on to say that the Commission's decision in this matter was contrary to the position of the United States at the 1959 ITU Conference. There it was agreed that caution in allocating portions of the spectrum was wise, for the process of displacing vested interests would be long and tedious. In fact, the United States had been the prime mover behind the allocation of frequencies for experiments in space.

At the Conference, the United States had proposed that ten bands be assigned to accommodate the functional aspects of space communications such as

---

[a]The frequencies involved were above 890 mc/s. The spectrum between 1000 and 10000 mc/s is considered the best for space services.

tracking telemetry, guidance, and command. The proposals sought to meet two criteria: (1) the best allocation from the point of view of propagation characteristics, and (2) the best allocation from the point of view of causing the least disruption of established services. However, while the criteria were met from the American point of view, other countries heavily used bands which were lightly used in the United States. Hence, the American proposal did not win complete acceptance. Thirteen narrow bands were allocated for research purposes; none were assigned on a primary basis.[b] The allocations were only for an interim period; but it was agreed that an Extraordinary Administrative Radio Conference would be called in 1963 to deal with the developing requirements for operational systems.

To prepare for the Extraordinary Conference, the FCC introduced Docket 13522, and IRAC continued its previous studies, including a long-range frequency allocation plan to accommodate the needs of all services.[c]

## Civil-Military Relations

A second early development in space communications involved the relationship between the Department of Defense and NASA, a subject which received much attention during the 1958 Congressional hearings leading to the establishment of NASA. The Space Act had provided for a division of responsibilities between the two bodies, but it became apparent that space communications could be a capability which would serve both military and civilian needs and that some of these needs overlap and others are distinct. Both sectors have a need for sending growing amounts of routine information overseas. But the military, and for that matter, the diplomatic requirements for secrecy call for more stringent controls over the communications channels than is the case with normal business traffic. National security requires a communications system which is secure, reliable, rapid, and capable of reaching any remote point on the earth. In the context of the "balance of terror," this capability should be present at all times. To the engineers, space communications offered a very promising means of meeting these needs.

By the initial division of labor between NASA and Defense, the civilian space agency was to sponsor the development of passive communications satellites and the military was to develop the active variety.[d] One rationale for this division of labor may have been to emphasize the civilian side of space communications. The technology to utilize a passive satellite was less complicated than that of the

---

[b]The following order of decision by the ITU is on a continuum of their authoritativeness: 1. primary 2. secondary 3. footnote 4. conference resolution 5. conference recommendation.

[c]See Chapter 5.

[d]Passive satellites reflect radio waves. Active satellites receive and transmit messages.

active system and NASA's predecessor agency, the National Advisory Committee on Aeronautics (NACA), had already begun work on this type of satellite. Hence, the passive satellite was thought to have greater promise of initial success than the active. Under this division of responsibilities, NASA successfully launched ECHO I on August 12, 1960. Within the Defense Department, the Advanced Research Projects Agency (ARPA) had been given the responsibility for satellite communications. It sponsored Project SCORE, a delayed repeater satellite, which was successfully orbited on December 18, 1958. This satellite was a space communications first, it being the first time a voice message was received from outer space. But the whole venture was a stunt to shore up American prestige, for the satellite did not have the capability to handle real-time communications. This was also true of Project COURIER, which achieved a successful launch on October 4, 1960. This project was managed by the Army, which had been given ARPA's duties by Neil McElroy, the Secretary of Defense.

By 1960, technological change altered expectations. Active satellites with real-time communications appeared to be just over the horizon. The Army had begun work on Project ADVENT, which was to be a 24-hour equatorial or synchronous satellite. NASA thus became interested in pursuing research on active satellites in order to fulfill civilian functions. Within the Department of Defense, on the other hand, the Air Force developed an interest in a passive communications project to be called Project West Ford.[e] Hence, the original division of labor proved unworkable. In August, 1960, a new understanding between NASA and Defense was reached, allowing each to pursue those interests it deemed worthwhile. "Unnecessary" duplication was to be avoided by assuring a close working relationship between the operating offices.

The swift pace of technological advance not only altered the original relationship between NASA and the Defense Department, but brought into focus the problem of how an operational system would meet the differing, yet in some ways similar, requirements of the civilian and military sectors. If the pattern that developed in the use of the submarine cables and land-lines were to be followed, the Department of Defense and the State Department as well, would lease channels from private carriers. Secrecy, where necessary, could be used for communications to remote regions of the world and for mobile communications. There would then be two interlocking national communication systems, one handling private and Governmental traffic, the other solely for Government use. The prospect of continuation of business as usual met an ambiguous response in the Senate and House Space Committees. In its December, 1960 report, *Policy Planning for Space Telecommunications*, the Senate Committee contended that the prospect of a dual role for communications satellites should not be overlooked.[9] Yet the report recognized that joint use of a satellite communications system might present problems in obtaining inter-

[e]See Chapter 6.

national cooperation. The report reiterated the intent of the Congress that NASA, rather than Defense, be the appropriate agency for dealing with other countries.[10] Almost a year later, in October 1961, the House Space Committee issued a report, *Commercial Applications of Space Communications Systems*.[f] Its emphasis was on the problems for our foreign relations of a dual-purpose system which could "conceivably be condemned by Soviet progaganda as 'militaristic,' and even neutral nations might object to the system's use for other than the most routine military purposes, making it more difficult to get international agreement on frequency allocations for space communications purposes, to negotiate for the establishment of ground facilities on foreign soil where that might be necessary, and to get needed cooperation of other nations on a variety of other matters essential to a viable system."[11] With this perspective, it was natural for the Report to favor continued work on a separate military system, but no explicit conclusion was reached to the effect that the Defense Department should forego participation in a civilian system.

The Air Force and the Navy favored a dual role for communications satellites.[12] This was in line with long-standing Government policy which provides that the Federal Government will not start or carry on any commercial industrial activity whose benefits could be procured from private enterprise. While this is the general policy, no authoritative decision had been made concerning a dual-purpose satellite system. The Senate Space Committee Report and the Report of the House Space Committee indicated that within Congress there was no crystallization of opinion on this issue. On the other hand, the Senate report had called into question all policy for space telecommunications and had criticized the adequacy of the policy-making machinery. It is to these criticisms that we will now turn. They are part of the continuing debate on this subject.

### The Adequacy of Policy and Policy-Making Machinery

The Senate report stated that the single most important issue facing Congress was the ratification of the treaty which grew out of the ITU Conference in 1959.[13] Most of its conclusions related to the problems of frequency management and the preparations for the Extraordinary Conference of 1963.[14] But, during the preparation of the report, it had become apparent that the problems were much more far reaching than those connected with frequency allocations. As witness to this concern the report expressed serious interest in the following questions and their interactions with each other:

1. The relationship between Government and industry
2. The image of the United States abroad

---

[f]This report was prepared by the Committee Counsel, Frank R. Hammill, Jr.

3. The satisfaction of the needs of the newer nations as well as those of our military and the civilian economy

The relationship between these questions is intimate, involving issues of both foreign and domestic policy. The Committee did not want the relationship between industry and the Government to develop in such a way as to lend support to the image which often develops abroad—that any United States program in space would simply enrich the private communications interests. In other words, the disposition of the question of ownership and regulation would affect our stature overseas.

The Committee report indicated concern not only with our image abroad, but with the concrete benefits which the Executive might seek to achieve by space communications, in the underdeveloped countries of the world. The question arose as to whether the Executive had a coordinated policy to harmonize these various goals and issues. The report stated:

The general problem of worldwide communications involves a complex and interrelated set of economic and political as well as technological considerations. Thus, any plans for such an important step as space service requires evaluation of broad national policies in the field of communications. At the present time, such policies do not appear clearly defined. Moreover, the mechanism for establishing policy is unclear.[15]

The report identified a myriad of operating and coordinating agencies,[16] and called on the Executive to identify a "central Federal authority for communications policy." You will recall that requests for a single communications agency had been issued before, but the pluralistic system continued despite these cries for central coordination. We will now see how the Executive responded to this new "challenge."

**Government-Industry Relations—1960-1961**

We have seen that the pace of technological innovation in satellite communications overtook the expectations of the policy-makers. The Congressional response to this predicament was the Report of the Senate Space Committee in December, 1960. Awareness of the Committee's ideas had seeped through the Executive before the report was officially published. The most immediate response came from Dr. T. Keith Glennan, the Administrator of NASA. On October 12, 1960, he proposed that satellite communications be left to private enterprise.[17] But, because launching and other expenses would be beyond the capabilities of private enterprise, Glennan proposed that NASA assist industry on a cost-reimbursable basis. Two months later, this general policy guideline was championed by President Eisenhower, who said:

This nation has traditionally followed a policy of conducting international telephone, telegraph, and other communications services through private enterprise subject to governmental regulation. We have achieved communications facilities second to none among the nations of the world. Accordingly, the Government should encourage private enterprise in the establishment and operation of satellite relays for revenue producing purposes.[18]

Before this speech, AT&T had made an "offer" to negotiate with the Government to establish specifications for a communications satellite. However, the Administration did not reply to this offer[19] but, rather, on January 4, 1961, it published a call for competitive proposals for the development of an experimental communications satellite system.[20] Certainly this action was in keeping with the antitrust laws. But, in any event, these actions by the Republican Administration at the end of 1960, indicate that it had become definite Government policy to encourage private ownership of space communications, in line with traditional communications policy. Traditional criticisms of this policy were not debated at this time, however. Problems of patents, subsidies, and monopolies at home and their effect on American prestige abroad lay dormant, but not for long.

On January 20, 1961, John F. Kennedy became President of the United States. In his State of the Union message on January 30, he invited "all nations—including the Soviet Union—to join with us in developing . . . a new communications satellite program. . . ." In conjunction with other cooperative efforts, the President hoped to end "the bitter and wasteful competition of the Cold War."[21]

Before his inauguration, the President-Elect had appointed an ad hoc committee on space to apprise him of the state of the national space effort. On January 12, this committee issued a report commonly referred to as the Weisner Report after the chairman, Dr. Jerome Weisner. This report analyzed the American space effort as a whole. To the ad hoc group there were five principal motivations of the space program: (1) national prestige, (2) national security, (3) scientific knowledge, (4) practical nonmilitary applications, and (5) international cooperation.[22] Communications capabilities were discussed under the last two objectives. In analyzing the feasibility of a radio and television space broadcast service, the report stated that "the development investment required is so large that it is beyond the financial resources of even our largest private industry." Thus, the report advised the establishment of "organizational machinery within the Government to administer an industry-government civilian space program."[23] No stand was taken on the private ownership issue, but support for some sort of close working relationship was certainly implied. President Kennedy accepted the Weisner Report leaving unanswered this question of public or private ownership. He was, however, opposed to the Eisenhower competitive bidding proposal whose actual outcome, he believed, would turn over satellite communications to the AT&T.[g] On the other hand, President

gInterview with Dr. Edward Welsh, October 12, 1965.

Kennedy did include the Eisenhower budget figure of $10 million in his supplemental budget request for NASA in Fiscal Year 1962. This sum would be used to work with industry which would then reimburse the Government.[24] A specific decision on the ownership issues would have to await the results of several months of Congressional hearings and deliberations within the Executive.

One of the first steps within the Executive was a working out of the relation between the FCC and NASA. On February 28, 1961, a memorandum of understanding was signed regarding the respective responsibilities of the two agencies.[25] Both agencies asserted that there was a strong feasibility of using communications satellites to satisfy the anticipated expansion of international communications. Beyond this, as a policy guideline, they affirmed the proposition that "the earliest practicable realization of a commercially operable communications satellite system was a national objective." And further, that "In accordance with the traditional policy of conducting international communications services through private enterprise subject to Governmental regulation, private enterprise should be encouraged to undertake development and utilization of satellite systems for public communications services." Both NASA and the FCC believed they had the statutory authority "to proceed expeditiously with the research and development activities necessary to achieve a commercially operable satellite system."

In investigating the form a commercial enterprise should take, the FCC was beset by avid interest on the part of the international common carriers. The problem was that these carriers envisaged a single or a limited number of systems. This opinion was based on the expense of such a system, the efficient use of the radio spectrum, and the likelihood that the existence of several systems would prove uneconomical. The question arose as to how to relate these technical and economic considerations to the antitrust laws and policies of the United States. To clarify this predicament, the FCC, on March 29, instituted an "Inquiry into the Administrative and Regulatory Problems Relating to the Authorization of Commercially Operable Space Communications Systems."[26] All interested parties were invited to communicate their views to the Commission. The first response of a Government department came on May 5, when the Department of Justice conveyed its views on the conditions that a commercial system would have to meet to be consistent with the antitrust laws. These were:

1. All interested communication common carriers to be given an opportunity to participate in the ownership of the system;
2. All interested communication common carriers to be given unrestricted use on nondiscriminatory terms of the facilities of the system whether or not they elect to participate in ownership;
3. All interested parties engaged in the production and sale of communication and related equipment be given an opportunity to participate in ownership of the system; and
4. All interested parties engaged in the production and sale of communi-

cation and related equipment be given unrestricted opportunity to furnish such equipment to the system whether or not they elect to participate in ownership.[27]

In effect, the Justice Department declared that the establishment of one system would be consistent with the antitrust laws if all the interested carriers and manufacturers could share in the rewards.

The FCC did not have the same views on the prospective system's composition. It believed that the established international carriers, by reason of their experience, were best suited to own the commercial satellite system.[28] Furthermore, the FCC contended that, under the Communications Act, the international carriers have the responsibility of providing efficient service, and, thus, they should be given the greatest degree of direct control of the proposed system. Accordingly, the Commission called a conference of the international common carriers which, in conjunction with certain Government agencies, would explore plans for establishing an operational communications satellite system.

On June 5, the conference met, although its participants included domestic carriers and manufacturers as well as the international carriers.[29] The diversity of opinions was such that no consensus was reached. AT&T and RCA Communications wished to restrict a new joint venture to the international carriers.[30] GT&E, a domestic carrier as well as a manufacturer, wished to limit ownership to international and domestic carriers.[31] And GE, a manufacturer, was interested in establishing a corporation made up of common carriers, manufacturing companies, and possibly the public.[32] Except for NASA, the FCC, and the Department of Justice, no other Government agencies had as yet crystallized their views at the time of the June 5 conference.

Due to the active interest in the business community and the differing views of the Department of Justice and the FCC, the matter of establishing one operational communications satellite system was pressed up to the Presidential level. On June 15, the Chief Executive asked the Vice President, as chairman of the NASC, "to make the necessary studies and Government-wide policy recommendations for bringing into optimum use at the earliest practicable time operational communications satellites."[33] In the body of the letter, the President expressed a firm interest not only in the form such a venture would take, but the policy objectives it should serve. He wanted to give particular attention to the communications needs "of this hemisphere and the newly developing nations throughout the world."

The President's emphasis on delineating specific goals was a reaffirmation of his speech to a Joint Session of Congress on May 25, 1961.[34] This speech is famous for its commitment of the Administration to the goal of putting a man on the moon by 1970. However, our attention is naturally drawn to the President's concern for space communications. He proposed making the most of

America's *present leadership* by accelerating the use of space satellites for worldwide communications. Speaking in general about the space program, the President said:

the facts of the matter are that we have never made the national decisions or marshalled the national resources required for leadership. We have never specified long-range goals on an urgent time schedule, or managed our resources and our time so as to insure fulfillment.

The President acknowledged the current Soviet lead in general, but obviously implied that in space communications, the United States was ahead. At the time of his speech, the United States had successfully launched three communication satellite experiments, while the Russians had probably launched none. Another implication of the President's remarks was that our lead in space communications was not the result of conscious, purposeful effort. Perhaps the President considered it to be an accidental offshoot of the reaction to Sputnik. He obviously intended to infuse new determination into the space communications area. One immediate step was a supplemental request for $50 million additional for NASA during Fiscal Year 1962. This brought the research and development figure for communications satellites to $94.6 million for that year.[35]

Another immediate step was to carry out the President's request to the National Aeronautics and Space Council. Dr. Edward Welsh, Executive Secretary of NASC, asked representatives of the following agencies to work out a draft policy statement: State, Defense, NASA, Atomic Energy Commission, Justice, FCC, the Office of Civil and Defense Mobilization, the Bureau of the Budget, and the Office of the President's Scientific Advisor.[36] As a result, a draft working paper was drawn up and circulated to the various agencies. On July 14, the Space Council held a formal meeting where unanimous agreement was reached on the recommendations to be sent to the President. The President evaluated these and issued a public policy statement on July 24.[37] The President invited "all nations to participate in a communication satellite system, in the interest of world peace and a close brotherhood among peoples throughout the world." He favored private ownership and operation of the United States portion of the system, provided that certain policy requirements were met. These included both foreign and domestic criteria. On the foreign policy level, he said the system must provide opportunities for foreign participation in the system and also be global in nature, even if this meant extending service to unprofitable areas. The domestic considerations involved assuring nondiscriminatory use and access to the system by all authorized carriers; and effective competition to be guaranteed by competitive bidding and the structure of ownership and control of the system. These goals were to be realized at the "earliest practicable date."

The Government would play a significant role in the launching of the system. NASA would continue its research and development activities, the fruits of which would be available to private enterprise. The State Department would conduct or maintain supervision of international agreements and negotiations. The FCC would regulate the system in fulfillment of its statutory responsibilities in the area of antitrust policy, frequency allocations, and rate charges. In addition, technical assistance to newly developing countries would spur the attainment of a truly global system. Coordination of these various activities would be the responsibility of NASC.

Concurrently with the President's statement, the FCC established an Ad Hoc Carrier Committee to consider the administrative and regulatory problems associated with a commercial space communications system. It is significant that this committee was to be restricted to international carriers, against the complaints of General Electric and General Telephone and Electric. Given the opinion of the most powerful of the international carriers, AT&T, that the composition of the new enterprise should be restricted to the international carriers, this action would seem to have prejudiced the outcome of the inquiries. However, the FCC contended that its participation in the committee meetings would assure objectivity. In addition, the Ad Hoc committee was enjoined to adhere to the President's policy objectives. The Antitrust Division of the Department of Justice was apparently contented with these safeguards, for it accepted this procedure.

While this movement toward concrete realization of a space communications capability proceeded within the Executive, committees in both the House and the Senate began holding hearings on the many issues involved. In the House, the Committee on Science and Astronautics commenced public hearings on May 8, 1961, which with interruptions, ran until August 10.[38] The day after the President's policy statement, the House Committee on Interstate and Foreign Commerce started hearings which lasted until July 28.[39] The Antitrust Subcommittee of the Committee on the Judiciary also held hearings in the House, but these touch space communications only tangentially.[40] In the Senate, the Subcommittee on Monopoly of the Select Committee on Small Business began hearings on August 2, and these lasted with interruptions until November 9.[41] These hearings provide a wealth of data with which to consider the attitudes of Business, the Executive, and Congress on the various issues which were listed at the beginning of the chapter. At this point, we will describe their attitudes on the ownership and regulations issue. This analysis will be built around the following question: What alternative forms could a commercial venture in space communications take, and how would these alternatives affect the realization of the President's foreign and domestic policy guidelines?

### Attitudes of Business

Industry had taken an interest in space communications long before the Russian launching of Sputnik. In 1954, Dr. John R. Pierce of Bell Telephone Laboratories, had suggested the use of satellites as a microwave relay.[42] In 1951, RCA had engaged in conceptual work,[43] and the Lockheed Aircraft Corporation had made its first study in 1954.[44] But it was not until the Government developed boosters capable of launching sizable payloads, that private enterprise began research on a continuous basis. Most of the early research activities were carried out by business under Government contracts with NASA and the Department of Defense. For instance, RCA supplied the radio equipment for Project SCORE.[45] But, by 1959, the pace of technological advance was such that it appeared to several companies that an operational system was just on the horizon. GE contended in 1959 that "the state and art in communications and astronautics both have reached the point where a useful, world-wide communications satellite system can be initiated immediately.[46] And Lockheed had concluded in the same year that the major problems were nontechnical; they involved legal, regulatory, economic, and international issues.[47] By 1960, AT&T expressed avid interest in getting a contract from NASA and envisioned a commercially operable system of about thirty low-orbit satellites by 1964.[48]

Industry's active interest in satellite communications did not produce a unified position on the form which a commercial venture in space should take. Most companies favored some sort of private enterprise, but even here opinion was not unanimous. The Philco Corporation considered that "at the proper time it will be desirable, and in the best interest of the United States, to turn control of some of these services and perhaps all of them, over to some international body, such as the United Nations, in order that they may be used most effectively to promote the welfare of the United States and the free world."[49] In 1959, GE had favored government operation initially, with movement towards commercial control as the system matured.[50] We have seen in connection with Docket 14024 that the alternatives on ownership were presented by representatives of three different types of business. International carriers tended to favor ownership by international carriers. Domestic carriers favored ownership by all carriers. And manufacturers favored ownership by carriers, manufacturers, and, possibly, the public.

In August, 1961, AT&T stated that the usefulness of a space communications system would depend upon its integration into the existing communications structure, which was controlled on the international level by the international carriers. To AT&T it seemed appropriate that the United States "should look to its common carrier industry for experience, competence, and financial ability to pioneer with satellite systems."[51] AT&T's Vice President and Chief Engineer, James E. Dingman, defended this position by asserting:

Hard as it may be for some to understand, our sole interest is in the earliest practicable establishment of a worldwide commercial satellite system useful to all international communications carriers and agencies here and abroad.[52]

The objective of usefulness to the international carriers is not too hard to believe, but it does contrast with the desire of Philco that the system be of use to the free world, and of President Kennedy who wanted the system to be of service to the entire globe. In essence, the AT&T position was that, if the

Government wanted an efficient system at the earliest possible time, it should leave matters in the hands of the international common carriers. If other interests were worried about AT&T's dominating the proposed system, AT&T suggested that they look to the regulatory powers of the FCC.

GT&E, a domestic carrier and a manufacturer, favored a system open to ownership by all common carriers who wished to participate.[53] The company contended that the system could meet domestic as well as international communications needs, and that ownership should be left in the hands of all common carriers who have already demonstrated their capability and know-how. Thus, GT&E took the same grounds as AT&T for ownership, i.e., experience.

Manufacturing companies like Lockheed tried to refute the place of experience as the primary reason for ownership. Lockheed favored private enterprise with participation by various private companies and the general public.[54] It feared domination by a single interest (AT&T). One reason Lockheed feared AT&T domination was its worry that Bell would go into the missile business to orbit its satellites for communications. To avoid this threat, Lockheed insisted that there be real competitive bidding for component parts of the system. However, to Lockheed, it was "extremely difficult to envision free competition in the hardware market for the system where the owners of the system are both operators and manufacturers."[55] Given this fear, it was natural for Lockheed to opt for a broad-based system.

In summarizing business attitudes, we can see that there was no dominant industry position. In general, the differing positions on the form of a private venture in space were related to the material interests of the companies—sometimes rationalized into a more exclusive interest. On the other hand, there was a case of a business, Philco, proposing a truly revolutionaly approach, perhaps in the general interest.

### Attitudes in the Executive Branch

From testimony at Congressional hearings by officials from the FCC,[h] the Department of Justice, NASA, the Department of State, the Department of Defense, and other participants in the decision-making process, there emerged a somewhat disoriented approach to the question of establishing one private company to carry on space communications services. We have seen that the FCC promoted a private venture wholeheartedly. An explanation of this attitude may lie in its statutory responsibilities. Being responsible for regulating private common carriers, the Commission proposed a form of ownership which would have enabled it to continue performing this task. At the Subcommittee on Monopoly hearings, the Chairman of the Committee, Senator Russell B. Long of Louisiana, asked Newton Minow, Chairman of the FCC, if he would prefer Government ownership if it could do a better job. Mr. Minow responded:

Yes, sir—of course, I would have to answer that in the framework of the Federal Communications Act—with my present position and statutory responsibility I

[h]Really an "independent," "regulatory" commission but operationally a participant in Executive Branch deliberations.

would have to answer it in terms of the policy and purposes of the Federal Communications Act—which imposes upon us the duty of trying to achieve and encourage a communications system under private enterprise.[56]

On the other hand, he went on to say that "if we felt that the present statutory setup did not enable us to operate or encourage a system in the public interest, we would let Congress know about it and hopefully make some recommendations." But he did not feel that way, stating: "the law presently is adequate."

The Department of Justice felt that if there were to be a private venture in space communications, it must adhere rigidly to the antitrust laws. In relation to its statement of May 5, 1961, in the proceedings of Docket 14024, the following statement finely presents the Department's view (or at least that of the Antitrust Division):

The Department of Justice firmly believes that a project so important to the national interest should not be owned or controlled by a single private organization irrespective of the extent to which such a system will be subject to governmental regulation.[57]

In the hearings before the House Committee on Interstate and Foreign Commerce, Mr. Loevinger of the Antitrust Division, contended that "regulation cannot eliminate the problems which would result from control of the system by a single company whether by ownership, by patents, or otherwise"[58] and to the Subcommittee on Monopoly, Mr. Loevinger made these far-ranging comments:

Satellite communications will by its very nature play an important role in international relations. The United States is engaged in a worldwide struggle to demonstrate that our economic system of free competitive enterprise can itself compete favorably with the Communist system of controlled monopoly. The satellite communications system can well be a prime example of the effective operation of the free enterprise system, and it is, therefore, of vital importance to the national interest that no single private concern dominate satellite communications.[59]

Mr. Loevinger was concerned that AT&T might gain dominance in the system through its relation with its manufacturing subsidiary, Western Electric, even though other carriers might be included in the venture.[60] Competitive bidding can be undermined by a common carrier if it uses specifications and standards which fit with its manufacturing subsidiary's interests, and this, in turn can significantly affect the pace of technological progress. Thus, for reasons of prestige and effectiveness and the responsibility to implement the antitrust laws, the Department of Justice favored a broad-based ownership in the prospective space communications system. And, alternatively, the Department thought that Government operation of the satellites as distinguished from ground stations, would not be inconsistent with the President's policy statement.[61]

James E. Webb, the Administrator of NASA, was less concerned with the question of ownership: "It is the development of a system that the Government is trying to bring into being. And the ownership of something out in space, I don't think, is the crucial question here."[62] Putting such emphasis on early development, Mr. Webb was inclined to give the international carriers, "already in business, doing this work," a first crack at planning the international system.[63] This apparent lack of concern for the ownership structure, which the Department of Justice thought would be of extreme importance to our foreign policy, may be traced to NASA's primary role as a research and development agency. It was beyond its normal jurisdiction to be concerned with the character of operating systems.[i]

The Department of State, surprisingly enough, did not have specific views concerning the form a communications satellite system should take. On July 17, Philip J. Farley, Special Assistant to the Secretary of State for Atomic Energy and Outer Space, stated:

Among the most perplexing problems . . . is that of determining the respective functions of the government and private industry in this country and the part to be played by other governments. Novel arrangements may be needed to deal with novel technology.[64]

With this introduction, one might have expected Mr. Farley to be concerned with the antitrust and monopoly aspects of the proposed system and how they might affect our image overseas; but no such analysis was forthcoming. He was very much concerned that whatever system ultimately was chosen should be global and flexible enough to serve the needs of small countries as well as large. The system should not only facilitate the linking of the United States with other countries but the linking of other countries among themselves. In addition, Mr. Farley thought that space communications could make significant contributions to our defense forces and the communications capabilities of our allies. Furthermore, as the coordinator of United States participation in ITU conferences, State hoped the proposed system would prove of benefit in managing the hardpressed frequency spectrum.[65]

The State Department saw the advantages of space communications in terms of their short term and long term consequences. It was not as concerned as NASA to establish a system as quickly as practicable "because of the importance of matching early availability with maximum usefulness."[66] The most immediate use envisioned was increasing "to an unprecedented degree the transaction of the world's governmental and commercial business."[67] Longer range benefits should include cultural exchange; more rapid dissemination of U.N. reports; service to U.N. emergency and peacekeeping operations; and contributions to worldwide meteorological activities.

The Department of Defense did not feel itself qualified to speak on the business organization suitable for a commercial system. However, Brigadier

---

[i]This is not true of the Apollo Program, but this is so unique as to be incomparable. Mr. Webb had long business experience, but perhaps his new role circumscribed his attitudes.

General William M. Thames, who was the commander of the Army ADVENT Management Agency, testified on August 9 that the ADVENT could easily be reengineered to provide a capability for handling all of the contemplated commercial and military traffic. Questions by Representatives Karth, Daddario, and Ryan pointed to the apparent duplication of effort in financing research in the military and civilian areas, if it was thought that one system could handle the projected demand.[68] General Thames was at pains to point out that there had been no duplication of effort in research and development, although he admitted there might be duplication in developing two operational systems, one for the military and one for commercial use.[69] We can see here that considerations other than economic determined policy in this area. A dual military-civilian system might not contribute to international cooperation due to the unwillingness of neutrals or enemies to join such a system. And, on the domestic front, such a system would undermine those interests which had heavy investments in cables and microwave relay stations. In addition, the development of a dual system would spur emotive debate on that highly political issue—"Big Government" versus "Private Enterprise." Perhaps it was worth the increased cost in dollars to avoid these political costs. But this formulation of the alternatives was not debated in the House Space Committee Hearings.[70]

## Congressional Attitudes

Congressional hearings were in the nature of investigations into the advantages and disadvantages of Government policies and organization for space communications. We have seen that the Senate Space Committee had called for an overhaul of the policy-making machinery and a clear statement of what the policy actually was. The new President had activated the dormant Space Council to assist him in policy-making, and, under the aegis of this body, a policy on space communications was presented to the Nation by the President on July 24, 1961. Therefore, it may be said that the Executive Branch had responded to the demands of the Senate Committee, although one must also mention self initiative and the advice of the Weisner Committee. But, in the summer and fall of 1961, three other Congressional committees held hearings in order to further survey the policy making process within the Executive.

In addition to Committee activity, there were of course many informal contacts and personal representations. One example is the August letter by three Senators, Humphrey, Kefauver, and Morse, and 32 members of the House to the President. They urged him to avoid making a hasty decision which they feared might give the international carriers a monopoly over space communications. They thought the resolution of the ownership issues should await the development of an operational system by the Government.[71] The letter indicated the attitudes of the liberal Democrats in Congress. To avoid a quick decision and wait until the Government had developed an operational system, would probably give the upper hand to those in Government who wanted a strict regulation of space communications in the general "national interest." The attitudes of the liberal Democrats, who might be called the nouveau Populists as far as this issue was concerned, were the natural response to the avid interest of AT&T in acquiring ownership for the international carriers. The dissatisfaction with this solution among many domestic common carriers, manufacturers, and those in the Congress and the Justice Department formed the rationale for the Congressional hearings.

The diversity of opinion was such that it was reflected in the attitudes of the Committees. The House Interstate and Foreign Commerce Committee, which had held hearings from July 25 through July 28, did not issue a report in the First Session of the 87th Congress. Rather, its hearings were a sounding board for the spectrum of opinion within the Executive branch. No industry representatives were heard, although business ideas were mentioned in connection with the FCC testimony on Docket No. 14024. Crystallization of the Committee's views would wait until further hearings were held in the Second Session.[j]

The House Committee on Science and Astronautics issued a report on October 11, 1961. The conclusions of this report were an example of the feeling that developments were so fluid at that time, that no final decision on ownership should be made. Rather, the Government should continue to play the major role in research and development; the decision on ownership should not delay this effort. The Committee was anxious that the Space Council should assume the leadership in coordinating policies which cut across the responsibilities of several Government agencies. As far as foreign policy was concerned, the report indicated special concern that the arrangements for the system reflect a good image abroad. In order to assure this posture, the report stated that "all negotiations with foreign nations concerning establishment and operation of any system be conducted by the Government, not by private corporations as in the past."[72]

The Subcommittee on Monopoly of the Select Committee on Small Business, which held hearings at intervals between August 2 and November 9, took a more partisan attitude on Government policy toward private ownership of the proposed space communications venture. This perspective was almost wholly the result of the views of the chairman, Senator Russell B. Long. He favored Government control of satellite communications, at least for the time being. He was unconvinced that the FCC or the President could regulate a private venture in the public interest. Senator Long was not opposed to a private enterprise system on principle, but he thought that the service should be made available to all and he doubted the willingness of a private venture to provide this capability.[73] He even favored asking the Russians to cooperate in order to cut the costs of establishing an operating system.[74] He was especially worried about

---

[j]See Chapter 4.

the effect of a private, profit-making system on our image overseas. Thus he asked:

Would it be better to say that there is a service being provided by the United States or that there is a Russian system provided for the people of the world and they do not use ours because the other one is owned by the American monopolies.[75]

Senator Long was not worried about American competitive private enterprise in space; what worried him was American monopoly in space. He found that the international carriers had not been effectively regulated by the FCC and therefore there was no reason to believe that the situation would be different with space communications.

These views of Senator Long were echoed by the testimony of Representative William Fitts Ryan[76] and Senator Hugh Scott.[77] The former introduced a concurrent resolution calling for Government ownership for two years in order to avoid premature decisions.[78] The latter considered international control to be the proper solution in order to "save the people of the world from the kind of competitive assaults which might increase rather than diminish tensions between nations."[79] These attitudes, and especially those of Senator Long, represented a less permissive approach to the question of private ownership than had been evidenced in the hearings of House Commerce and Space Committees. As one would expect, the spectrum of opinion in Congress was as multifaceted as in industry and the Executive. The alternatives on the ownership issues ranged from favoring the continuation of control over international communications in the hands of the international carriers to favoring some sort of international control.

The dominant weight of opinion at this stage of the decision-making process favored a private venture with Government regulation, if competitive bidding and nondiscriminatory access to the system could be assured. Thus competition would not be promoted through establishing several communications satellite systems, but by establishing a joint venture where no one company would dominate and by FCC regulation. If possible, such a system would live up to the guidelines in the President's policy statement of July 24, 1961, at least insofar as domestic policy was concerned. What about the promotion of foreign policy goals? Here, all business interests wanted America to establish the system at the earliest practicable time for reasons of prestige and national security.[80] The fear of a Russian operational first in space communications was also a constant motive in the thinking within the Executive branch and in Congress. The Russian challenge, real or imagined, was a prime stimulus to rapid development of a space communications system. We have seen that the attitude to hurry up did not act as a damper on debate on the form such a system should take. But the push towards rapid development also raised other issues. How rapid was development to be? The answer to this question depended to a certain degree upon the state of development of alternative technologies. Here there were disagreements concerning the best type of system to achieve an early capability and also meet the President's other policy objectives: global coverage and efficient service throughout the world, including the less developed areas. It is thus fitting that we now turn to a consideration of the technological characteristics of the proposed systems as they were viewed by the various participants.

## Technical Characteristics of Alternative Communications Satellite Systems

To deploy a communications satellite system, it is necessary to have three segments: (1) a rocket launching capability, (2) communications satellites, and (3) ground stations or antennas to transmit and receive messages. In this section, the technical characteristics of communications satellites will be emphasized. We are concerned with communications satellites which facilitate communications between points on the earth, not with all satellites with communications potential.

Communications satellites can be classified as to whether they are active or passive, and, if active, as to whether they are delayed repeater types or real-time repeaters. Passive satellites merely reflect radio beams, while active satellites amplify and retransmit signals. Delayed repeaters receive a message from one point and on command transmit the recorded message to another point. Active satellites instantaneously relay the signals to properly equipped ground receivers. By the summer of 1961, the Government had launched three communications satellites, one passive and two delayed repeaters. For both commercial and military-diplomatic use, however, active, real-time repeaters were envisaged.

Communications satellites may orbit in essentially two types of orbits: (1) low or medium and (2) 24-hour equatorial. The latter orbit is at approximately 22,300 miles, and, if it lies directly over the equator, will remain, or appear to remain, stationary over a specific point on the earth because its speed will be the same as the earth's rotation. Thus it is called a synchronous satellite. Satellites in low or medium orbits, on the other hand, will be constantly moving with respect to points on the earth. These two orbiting possibilities affect several other characteristics of the system. A low-orbiting system requires two sets of mobile antennas, one to track one satellite as it moves across the horizon, and another to pick up the next satellite as it appears on the rim of the horizon. It would require from twenty to fifty satellites to cover the globe, depending on the altitude of the orbit. In comparison to a synchronous system, a lower system would require less booster power to put into orbit satellites of comparable weight, although it would also require more rockets, unless clusters of satellites could be put into orbit at one time. A synchronous system requires only three satellites to cover the whole globe except for remote polar regions. On the other hand, very sophisticated equipment to control position and altitude of the satellites is required. Given these differing characteristics of synchronous and random-orbiting satellites, let us now examine the attitudes of the participants on the question of choice for the system most likely to achieve President Kennedy's goals of rapid development of a global system.

Most observers believed that the low-orbit system would be easier to establish at an earlier date than the synchronous system.[81] On May 18, 1961, NASA had awarded a contract to have RCA develop such a system, called RELAY, which would move in orbits 1,000 to 3,000 miles above the earth.[82] AT&T had lost in the competition, but, at its own expense, started working with NASA on July 27 to launch two experimental satellites during 1962.[83] This project became known as TELSTAR. While RCA itself was working on a low-orbit satellite, it considered the synchronous system to be the wave of the future. Such a system would be more flexible and have a greater capability. "The flexibility of this approach is such that each country may have its own terminal facilities in its own territory, avoiding any need for retransmission from a centrally located ground station situated in or beyond other national areas."[84] Furthermore, the RCA representative asserted that:

The ultimate advantages of a synchronous satellite system raise a practical economic question as to how much should be invested by industry in a low altitude commercial system primarily to gain a certain amount of time. This is further complicated by the fact that the pressures to pay a premium for time are based more on national, political, and psychological factors than on commercial considerations.[85]

Thus, for RCA, the motivation for rapid development of an operational system was political rather than economic. Satellite communications would provide a capacity twenty to thirty times greater than the present transatlantic submarine cables; but this capacity would be needed in the future, not immediately.[86]

For certain other companies the prime motivation for developing an early capability was also political, but they contended that a synchronous system could be developed earlier. GT&E, which through its subsidiary Sylvania, had been working on ADVENT, said that the advanced system could be put up faster than the inferior low-orbit type.[87] The company also asserted that the random orbit system would not give global coverage and would need more expensive ground stations, thus deleteriously affecting communications to the less developed countries.[88] Thus, this system would not meet all the President's policy objectives. "A random system could discredit us before the world as a leader in space communications if Russia establishes a stationary satellite system."[89] The Lockheed Aircraft Corporation stated that "although one frequently hears that a low altitude system can become operational much sooner than a high altitude system, this, in fact, may not be the case."[90] The most optimistic attitude on the early prospects of the synchronous system came from the Hughes Aircraft Corporation. On May 25, 1961, the company had stated that it could put a three hundred voice channel satellite in synchronous orbit within a matter of months.[91] The cost would only be one-fifth the cost of other estimates. Whether Hughes could have delivered on this prediction, we will never know, for, at this time, NASA was planning for an early launching of RELAY and TELSTAR experiments. On August 11, however, NASA awarded Hughes a contract for a synchronous satellite called SYNCOM, with a launch target date of late 1962, but its capability would be only one voice channel.[92] In November, Hughes was of the opinion that a worldwide TV service could be established within one year if developments were organized properly.[93] Thus, leaving margin for considerable error, the prospects of fruitful development of a synchronous system appeared so favorable that many companies were seeking early establishment of such a system. On the other hand most observers thought the low-orbit system had the best *early* prospects. Each camp defended its position with political as well as economic and technical opinions.

In any event, the expectation for an early capability appeared so bright that the Government was pressed to decide how a communications satellite system would fit into the existing communications network. Would the new capability be integrated harmoniously into the existing system or would it transcend it and make it obsolescent? This question leads us into a consideration of two additional questions: (1) Did satellite communications constitute a revolutionary breakthrough? and (2) How would projected demand for communications services be met by satellite as compared with more traditional modes of communications?

### The Question of Technological Breakthrough

There were essentially two views of whether satellite communications were a technological breakthrough, and these views were related to the stakes of communications carriers and hardware companies in the traditional equipment. Two international carriers, AT&T and ITT, saw communications satellites as simply another type of long-distance communications device. To Mr. Dingman of AT&T "the proposed system would supplement and expend existing overseas communications facilities."[94] To Dr. Busignies of ITT the satellite "is like the present cables or microwave relay stations, it is merely a transmission element."[95] Since it was only a new device, but not a revolutionary one, these international carriers contended that they should be given control over satellites with communications capabilities just as they had control over other means of communication.

Several manufacturers, GE, Philco, and Bendix conceived of communications satellites as a revolutionary breakthrough. Perhaps the most imaginative comments were made by David B. Smith, a vice president of Philco. He said:

We think this technical achievement . . . is far greater than just the extension of our communication ability . . . But if we label that achievement as just being an extension of our present commercial communications service, certainly that is the way the rest of the world will regard it . . . We would urge that the Government establish a grand strategy to utilize these achievements to lessen world conflicts and to ease world tensions.[96]

Given this position, it was natural for Philco to favor international control over satellite communications rather than control by the international carriers. Seeing space communications as revolutionary, it favored a revolutionary type of organization to operate a satellite communications system. On the other hand, Mr. Smith recognized that if it were not considered to be revolutionary it might not have revolutionary consequences. GE and Bendix did not go so far as Philco. They favored ownership by private enterprise, but were unwilling to restrict the composition of the owners to the international carriers. To support their view, they argued that the technology was unique.

Within the Government, the tendency was to see space communications technology as novel, but not revolutionary. Commissioner Craven of the FCC regarded it "primarily as another means of relaying long-distance communication."[97] On the other hand, he saw the effects as revolutionary "in terms of national and international communications propaganda."[98] The Commissioner's answer to our original question would thus be more sophisticated than just a yes or no answer. One can see the capability as just another addition to the communications network and, at the same time, see its effects as revolutionary. The Senate Space Committee's *Communications Satellite Report* took this position when it said that the new technology was essentially "another type of long-distance, trunkline communications circuit,"[99] but also called space communications "an entirely new concept of worldwide communication . . . that would afford a unique potential as an instrument of international affairs . . . "[100]

Viewed precisely, one could say that the technological innovation in space communications was not a revolutionary breakthrough but an engineering application of existing know-how to the space environment made possible by the breakthrough in rocket technology. On the other hand, the effects of the engineering application could be foreseen as far-reaching, and perhaps, revolutionary. Viewed from the technical standpoint, the global system the President desired would be a composite of present facilities augmented by new cables and satellites. The relative place of the different techniques would depend on their reliability and capability in the face of future demands. Viewed from the economic and political angle, however, the structure of a new communications network could be consciously manipulated in such a way as to serve either inclusive or exclusive interests. Here is where the real breakthrough could occur. And this possibility affected the expectations of a great many important people. We have examined some of these expectations. Now let us look at the projected demands for communications services to further clarify the overall picture.

## Demands for Communications Satellite Services

In estimating the demand for communications satellites, one must distinguish between the several types of service such satellites could provide. These services are: telephone, telegraph, teletypewriter, telex, datafacsimile, and television. These services could be used singly or in combination for various different purposes—military, diplomatic, commercial, press, and to serve international organizations such as the World Meterological Organization and the International Civil Aviation Organization. In the early 1960s realistic estimates of future demand would have to be inferred from the growth rates of established services, and these were the telephone and the telegraph.

Instantaneous broadcast of television on a global basis was a future medium of international communication, so that its impact could not be analyzed merely in terms of rates, hours in use, and total words or images conveyed, but would have to include examination of audiences, the nature of the program, and the differences between telephone, radio, and television in their psychological perspective.[101] But while demand for TV could not be precisely projected, there was speculation about its impact. Some participants saw it as a way to project our civilization and as a means of producing a universal language. Others were less optimistic, seeing the different time zones and the differing languages as barriers to an international television service.[102] Thus, some participants saw television transforming the international environment and others saw it being engulfed by the same environment. Being a very risky matter, direct broadcast TV was not the motivating factor behind the avid interests of AT&T and other companies. A prime motivating factor here was the prospect that satellite communications could be used to ease the rising demand for overseas telephone service.

Since 1927, there has been a constant increase in the use of the telephone for international communications. Dr. Henri Busignies of ITT told the House Space Committee in May, 1961 that a 15 percent yearly increase in transatlantic telephone calls was the norm, and that with the introduction of expanded and reliable capabilities, the demand for service was apt to grow. "For example, the introduction of the first Atlantic cable in 1956 caused a 90 percent increase in the number of calls."[103] With increases in existing cable capacity, Dr. Busignies estimated there would be a total cable channel capacity of 1,500 in 1970 and that we needed to meet requirements of 4,000 to 6,000 by 1980.

Mr. Dingman of AT&T said in testimony before the same Committee that overseas telephone calls were "expected to increase from 4 million in 1960 to 20 million in 1970 and nearly 100 million in 1980. This means that overseas telephone circuits will have to be increased from the some 550 we have today to about 12,000 in 1980."[104] The Ad Hoc Carrier Committee also made an estimate of channel requirements for future years but added estimates of density for particular areas. Table 3-1 indicates the present and future voice channel requirements as seen by the international carriers in 1961.[105]

**Table 3-1**
**Projected Needs for Voice Channels**

| Areas | 1970 | 1975 | 1980 |
|---|---|---|---|
| Europe | 1400 | 2400 | 4300 |
| Africa | 70 | 150 | 200 |
| Near East | 80 | 100 | 150 |
| West Indies | 1200 | 2000 | 3300 |
| Central America | 150 | 250 | 450 |
| South America | 450 | 950 | 1800 |
| Alaska | 350 | 650 | 1300 |
| Hawaii | 500 | 850 | 1500 |
| Asia and other Pacific Areas | 450 | 650 | 1000 |
| Total | 4650 | 8000 | 14000 |

An interesting aspect of this table is that it points to the regional requirements of communications traffic. The principal variable determining density of communications is the flow of trade and business, except during war. This was proven by the Communications Policy Board in its 1951 Report to the President and is still true today.[106] The flow of messages tends to follow the flow of trade, 85 percent of overseas messages being commercial. Thus, although the Ad Hoc Carrier Committee envisaged a global system, the allocation of communications, at least of a traditional type, was estimated to fall in those regions whose commercial relations with other regions were advanced. Most participants in the Congressional hearings, when they referred to increases in international communications, used projected increases in the North Atlantic area as the prime example. One may conclude, therefore, that although a global system was envisaged, some parts of the globe would be more equal than others. The immediate uses foreseen by industry were, as Mr. Farley of the State Department said, to facilitate "to an unprecedented degree the transaction of the world's governmental and commercial business."

What were the uses that the les well-developed countries might make of space communications? The Senate Space Committee *Policy Planning Report* had pointed out that these countries were primarily in need of local rather than international traffic services. Nevertheless, the report indicated that many former colonies must depend for their international communications upon relays through their previous "mother countries:"

These new nations are certain to desire independent communications systems as part of their programs of self-government. A single space relay over Africa has entertaining possibilities as a link between all African nations and may prove attractive regardless of purely economic considerations. The United States may thus have a unique opportunity to assist these nations in their goals of self-realization.[107]

Let us now characterize the gestalt of the decision-making process as it appeared at the end of 1961—before the start of the second session of the 87th Congress which was to pass the Communications Satellite Act.

## Characterization of One State of the Decision-Making Process

The year 1961 came to an end with two separate events whose characteristics were to play a considerable part in the bargaining leading to the establishment of the Communications Satellite Corporation. One of these was the plan of the Ad Hoc Carrier Committee[108] and the other the address of Adlai Stevenson at the United Nations on December 4.[109] The ramifications of these two inputs into the policy-making process were not harmonious.

The Ad Hoc Carrier Report advocated the setting up of a nonprofit management corporation to *develop, operate* and *manage* the proposed satellite communications system. This corporation would act as a common carrier's common carrier, leasing circuits only to authorized carriers who would *own* the ground stations and the satellites. The board of directors of the corporation would be composed of representatives of the carriers plus three public directors appointed by the President.

To evaluate this report, the Senate Subcommittee on Monopoly held hearings on November 8 and 9. The reaction of the subcommittee can be indicated by this comment of its chairman, Senator Long:

There may be a way of giving it directly to AT&T, but short of that, it seems to me you could not have made a better recommendation to achieve the same result.[110]

The gist of the Senator's analysis was that the carrier proposal would involve fitting space communications into the existing system, thus increasing its cost to the taxpayers and decreasing the chance that it would be available to everyone on an equitable basis. The fact that the proposed corporation would not own the system but only manage it made it a screen for the interests of the carriers.[111]

In the course of the hearings of the Monopoly Subcommittee, it became evident that all the members of the Ad Hoc Committee itself were not committed wholeheartedly to the report. Western Union was worried that AT&T might dominate the endeavor.[112] And RCA Communications was not committed to any particular form of ownership and organization.[113] In addition, within the Government the Antitrust Division of the Justice Department was still deeply concerned about the monopoly aspects of the international carriers position in the proposed corporation.[114] Mr. Loevinger contended that the proposal was vague. He attributed these inadequacies to the effort of the participants to gain the widest possible agreement.[115] In his opinion, the issues of ownership and regulation were so important that Cabinet level consideration was an urgent requirement.

From this we can see that the report of the Ad Hoc Carrier Committee proposed a solution which, although it jelled with past practice, received lukewarm support in some quarters and outright opposition in others. The necessity for higher level consideration was thus apparent. In fact, this was already occurring within the Executive branch. Dr. Welsh of the Space Council was chairing a group which would offer suggestions to the President in December 1961. These would take the form of a legislative proposal in the second session of Congress.

The second event which brought 1961 to an end, and the consideration of space communications to a new beginning, was the address of Adlai Stevenson to the United Nations in December. In this speech, the Ambassador elaborated on the September speech by President Kennedy before the General Assembly. The President had called for cooperative efforts to establish "a global system of communications satellites linking the whole world . . . "[116] Ambassador Stevenson reiterated this call and warned the delegates on the relationship between scientific and social change:

Unless we act soon the space age—like the naval age, like the air age, and the atomic age will see waste and danger beyond description as a result of mankind's inability to exploit his technical advances in a rational social framework.[117]

Mr. Stevenson contended that "we are conditioned to think in terms of nations." He asserted that "Such concepts have no meaningful application to the unexplored, unbounded, and possibly unpopulated reaches of outer space." He then proposed a five-part program for cooperation in outer space. One part was the espousal of the desirability for a global system of communications satellites.[118] Ambassador Stevenson said that "this fundamental breakthrough in communication could affect the lives of people everywhere."

It could forge new bonds of mutual knowledge and understanding between nations.

It could offer a powerful tool to improve literacy and education in developing areas.

It could support world weather services by speedy transmittal of data.

It could enable leaders of nations to talk face to face on a convenient and reliable basis.

The enumeration of these goals and hopes "obligated" the United States before the "Town Meeting of the World" to promote the establishment of a space communications system which could best lead to their fulfillment. This far-reaching program differed from the plan of the international carriers for promoting their own control over the new system. The carriers were interested in carrier ownership: our U.N. delegate in ownership by interested members of the United Nations. In 1962, these two positions were at extremes in the debate over ownership and control. Before examining this debate, it will be fitting to conclude this chapter with some tentative findings on the three themes of this book.

## Conclusions

### The Influence of Technological Changes on Policy and the Policy-Making Process

In order to understand the effects of technological change on the rate of change and the relationship of the established institutions, one must first decide what a "breakthrough" is and how it differs from an "engineering application." This decision, as we have seen, can itself be a political decision beyond the bounds of purely objective considerations. The answer will depend upon the uses one foresees for the innovation.

Whatever the precise character of a technological change in space communications was from a technical point of view, it is obvious that there was considerable prospect of change from a political and economic point of view. The new capability promised to change the size if not the shape of the structure of international communications. Thus, there was an early responsibility for the Government to set policy in this area in which some agencies and departments had statutory commitments. It was natural, given the number of interests affected, that the President's policy statement of July 24 should include all plausible goals, but not state them in any order of priority. It was politically feasible to identify all the multiple values, although it might not prove operationally feasible as an outcome of the policy process.

Another effect of technological change was to create a new policy issue with new relationships among policy-making participants in an area which developed overlapping jurisdictions. The FCC was responsible for communications and NASA for space activities. There was early coordination between them on a basis of mutual adjustment and this coordination was enhanced by the operations of NASC, which had been activated by President Kennedy on the advice of the Weisner Committee. The attempt at coordination went part way toward meeting the demands of certain elements in Congress, but, in its very nature, this initial coordination was not complete because it related to problems which were the concern of other participants as well. Thus the push for an early capability by NASA conflicted with the antitrust policies as perceived by the Justice Department.

With its responsibilities for foreign policy formulation, the State Department entered the policy-making arena. It was more interested in enunciating general policy goals than working on the concrete aspects of an operational system. In working together, these various agencies proceeded on a basis of sequential mutual adjustment. There was no real central coordination except at the general policy level.

The pace of change had its effects on the policy making process. The vision of rewards tempted each interest to gets its foot in the door. The time for careful consideration by the Executive and Congress was scarce. An increased technical ability, which had not been expected to appear so quickly, constricted the time available for reconnaissance. Adding to the sense of urgency, however, was the motive to beat the Russians. Thus, we see a combination of technological change and a political goal increasing the sense of urgency. To entirely isolate either variable is impossible, although one can say that the realization of a technological change in space communications came first.

*The Clarity of the Goals of U.S.*
*Space Communications Policy*

We can see that in the space communications field, issues of foreign policy and domestic policy overlapped. The ownership of the system was thought to reflect on our image abroad as well as on the efficiency with which an operational system would provide services. In order to gain a perspective on the various issues[k] as they appeared in 1961, I will place them in three categories: domestic, a middle category for the "penetrated system" where resolution consequences are both foreign and domestic, and foreign.[119] The distinction between foreign and domestic policy is defined as the degree to which the rewards of domestic groups are affected by the outcome. Given this definition, it is easy to see why a middle category is needed to reflect the fact that some issues have both foreign and domestic consequences. The rationale for putting issues in one category or another is my subjective appraisal of their character.

Foreign Policy Issues:
1. Peace
2. Understanding
3. Global coverage
4. Service to less developed countries
5. One system
6. Participation by foreign countries
7. Frequency allocations
8. United States leadership

[k]For the purposes of this discussion, an issue is an area of intellectual, emotional, ideological, or structural behavior about which there is disagreement as to whether it exists, did exist, will exist, or should exist.

Penetrated system issues:
1. Urgency
2. System choice
3. Regulation of rates
4. Supervisory role of Government
5. Ownership

Domestic policy issues:
1. Government-industry cooperation in research and development—past, present, and future: the subsidy issue
2. Ground station ownership
3. Competitive bidding
4. Equitable access
5. Patents

These issues were given varying degrees of attention by the participants at different stages in the policy-making process. Throughout 1961 debate was focused on the second and third categories. The first-category issues were not the subject of great activity. The one exception is the question of frequency allocations. The other seven issues were either agreed upon or ignored. Their specific meanings were never questioned, yet the outside observer can identify several possible conflicts. Two examples will suffice. First, United States leadership might not be compatible with increasing understanding and participation by other countries. Other countries might want to participate as equals. In the second instance, one may see that increased international communications of the mass media variety might not aid in the orderly development of the less developed countries but contribute to anarchy and frustration. Expectations may rise only to be frustrated, if new means of communications are not used in a judicious manner.

We may conclude, then, that at this stage, the goals of American foreign policy for satellite communications lacked preciseness. This legitimated the activities of practically every participant in the policy-making process and gave President Kennedy the potential of receiving the political support of diverse interests. Furthermore, the abstract quality of the goals might be more fitting for an early stage of the policy-making process than an explicit ordering according to some a priori scale of values. The latter procedure would have hardened thinking and not given the opportunity for mutual adjustment in the light of future contingencies. This line of reasoning supports the arguments for pragmatism and incremental rationality. Sometimes "muddling through" may be more appropriate than cost effectiveness, especially for decision making on a complex public policy problem involving many intangibles and nonmeasurable variables.

The multiplicity of participants and possible policy goals point to the legislative character of the policy-making process. What we see is a problem area whose interests transcend and transform the traditional communications environment.[120] In early 1961, it was the expectation of the FCC, AT&T, and other international carriers that ownership should naturally fall within the boundaries of the established structure of international communications. If space communications had been viewed as a technique similar to the 1956 transatlantic telephone cable (TAT-1), no doubt this expectation would have proved true. But the Justice Department and other companies, particularly manufacturing concerns, objected to the FCC-AT&T position to which NASA was a technical partner.

# 4

## The Passage of the Communications Satellite Act of 1962

### Drafting of the Legislative Proposal Within the Executive Branch

The story can begin with President Kennedy's reaction to former President Eisenhower's proposals for Fiscal Year 1962. It will be recalled that the figure of $10 million for NASA, originally set by Eisenhower, was accepted by President Kennedy in his supplemental budget request.[a] However, the President disagreed at this time with the Eisenhower policy of giving private enterprise the responsibility for developing satellite communications. The President thought that this policy in reality meant turning control of space communications over to AT&T.[b] Instead he asked his advisors to recommend a policy which would avoid the pitfalls of monopoly control. This resulted in the President's policy statement of July 24, 1961.

### The Interdepartmental Drafting Committee (The Welsh Committee)

Shortly after President Kennedy set policy on an operational system for space communications, he asked Dr. Edward C. Welsh to make recommendations on the best means of implementing the policy. The President was especially concerned that an operational system be established at the earliest practicable time. Uppermost in his mind was the desire to beat the Russians in this new endeavor in space.[1] The memory of Sputnik was a powerful stimulus to this motive. The President did not request a legislative proposal, but Dr. Welsh considered this path the most efficacious. The politics of the issue had become so intense that resolution of the matter by Congress was necessary. This would insure the widest possible participation in the eventual outcome. This decision itself is notable from the point of view of the interaction between foreign and domestic policy in the penetrated system.

Dr. Welsh established an interdepartmental committee consisting of the same agencies and departments which had drafted the policy statement of July 24. This committee met frequently during the fall of 1961. It considered three alternatives with respect to ownership: (1) Governmental ownership, (2) carrier

---

[a]See Chapter 3.

[b]Interview, Dr. Edward Welsh, October 12, 1965.

ownership, and (3) private, broad-based ownership. These options were not examined according to their technical and organizational consequences alone, but, of necessity, were filtered through the political and psychological climate of opinion. This ethos concerned certain vague, but highly emotional issues such as "creeping socialism" and "predatory monopolies." Political as well as organizational feasibility had to be considered.[2] If one agency considered that the system should be owned by the Government, it knew that the public airing of this view would incur the wrath of those who were ideologically committed to free, competitive private enterprise. If another agency favored ownership by the international carriers, it knew that proposing this alternative would meet the bitter attacks of those wedded to a firm construction of the antitrust laws. In short, the issue was political; it involved polemics and emotions as well as dispassionate consideration of the pros and cons.

At one extreme during the interdepartmental meetings was the Justice Department which favored Government ownership, because competition could not really work if there were to be only one system.[c] At the other extreme was the FCC which favored continuation of the status quo by providing for the integration of the new technology into the existing communications setup, because that is where the experience and expertise lay. Early in the meetings, the Justice Department was persuaded that the only realistic alternative was private broad-based ownership. This was the dominant position, not because it was the only workable solution, but because it was the only solution Congress would accept. From the point of view of group politics, it is interesting to note that this position was not a compromise resulting from the opposition of equally powerful economic interests representing carriers and manufacturers. Rather, it was a compromise which reflected these economic interests and the political interests associated with the Justice Department and segments in Congress with a firm commitment to the antitrust laws.[d] These political interests had no economic stake in the outcome but were motivated by their conceptions of the public interest. In the case of monopoly, this public interest was seen as the promotion of individual opportunity through the prohibition of economic concentration.

## The Role of the Department of State

The representative of the Department of State was in a quandary with regard to the ownership problem. The State Department was internally divided. The Offices of International Organization Affairs and the Legal Advisor favored Government ownership, while the Bureau of Economic Affairs and the Office of

[c]This was Katzenbach's initial position according to Lee Loevinger; interview, October 26, 1965. This was also Dr. Welsh's initial position according to Ralph Clark; interview, July 26, 1966.

[d]See August 24 letter by antitrust segments in Congress, pp. 29-30; discussed in Chapter 3.

International Scientific Affairs (now International Scientific and Technological Affairs) favored a broad-based ownership under private enterprise. The representative of State to the interdepartmental committee thus had to represent two diverging positions. On the other hand, the Department was unified in its support of the foreign policy objectives a communications satellite system should serve. These were:

1. The United States should develop communications satellite technology for the benefit of all countries.
2. The communications satellite system should be organized internationally and not as a purely domestic corporation seeking profits.
3. The benefits of the system should be extended to remote and underdeveloped areas even though this might be unprofitable.
4. Other countries should be guaranteed the opportunity to participate in the established system, although, as a practical matter, the system should start on a small scale in the North Atlantic area, due to economic and technical considerations.
5. Whatever the outcome of the ownership debate, there should be enough Government control to assure that the above objectives are realized.[e]

It can be seen that these goals paralleled those of the President in his policy statement of July 24, 1961. In addition, the President had favored private ownership *provided that* this could be set up in such a way as to meet his policy requirements. Within the Department of State there were strong feelings that the President's goals could not be met by a private corporation. In order to solve the dispute within the Department so that a united front could be presented to the Congress, the Under Secretary of State, George W. Ball, requested that alternative draft memoranda be submitted to him. The Under Secretary chose to support the view that the policy objectives could be met by the formation of a private corporation under Government regulation. This decision then became the official position of the Department.

The fact that State recognized the foreign policy implications can be seen in its defense of language in the bill which would enable it to conduct or supervise foreign negotiations of the corporation. From this, one can infer that the State Department foresaw the possibility that a private corporation controlling satellite communications might not always act in the best interests of the United States. On the other hand, it did not go so far as to argue that Government ownership would increase the chances of foreign participation in the system. As most foreign countries have government owned telecommunications systems, this alternative might have received more consideration than it did, had it not been for the domestic climate of opinion which favored private ownership.

[e]Interview, Mr. Robert Packard, The Office of International Scientific and Technological Affairs, Department of State, April 9, 1965.

Instead, the Department's final position on the ownership issue reflected the weight of opinion within the United States.

From another vantage point, one could defend State's attitude as reflecting a long-thought-out position. With a firm background of cooperation with the international carriers which had been able to establish the international telephone and telegraph arrangements with foreign governments, it would have seemed anomalous to create a novel arrangement to handle space communications. This was the view within the Bureau of Economic Affairs. The trouble with this position is that it neglected the fact that space communications was envisioned by the President as serving more spectacular purposes than those of the traditional means of communications. A novel technology serving novel purposes might just require novel arrangements.

## The Culmination of the Work of the Welsh Committee

The discussions within the interdepartmental committee were mostly concerned with how to set up one company which would not bring with it the disadvantages of monopoly. The primary means of achieving this goal was believed to be the establishment of a system of checks and balances within the company itself. If there were to be no competition externally, there would be internally. Thus, the board of directors would have representatives of various carriers and the general public. Externally, the public interest would be guarded by specifying in detail the powers of the President, the FCC, NASA, and the Department of State. Thus, in essence, there would be two sorts of control on the corporation: (1) internal controls due to the diversity of views to be represented in the board of directors and (2) external controls lodged with different Executive entities. One set of controls would work almost automatically and the other consciously, to achieve the purposes of the legislation.

The Justice Department, represented by Mr. Loevinger, thought that external regulation alone would not be sufficient. He stated the rationale thus:

... What we are doing is trying to achieve many of the benefits and objectives of competition by virtue of the automatic operation of the company through its internal structure. And we believe that automatic operation in general is to be preferred to an attempt to secure economic performance by Government regulation.[3]

During the subsequent Congressional testimony, it became clear that the FCC objected to this system of internal control proposed by NASC and the Department of Justice. The commissioners felt that ownership should be restricted to the common carriers and that abuses should be prevented by allowing for specific safeguards. The pros of carrier ownership as envisaged by the FCC were: (1) the carriers will be more responsive to the communications needs of the public; (2) the new system could be more easily integrated into the existing worldwide communications network; (3) arrangements with foreign governments and entities would be simplified; and (4) carrier ownership would expedite maximum use of the system on a worldwide scale.[4] The common foundation of these assertions was that the carriers were experienced in international communications. A new corporation would merely be a carrier's carrier. It would have no direct dealings with the public and thus would blur the lines of statutory responsibility between the carriers and the public. In addition, the FCC thought that a new corporation would incur higher costs during its initial operational years and this, in turn, would inhibit its use by the established carriers.[5]

The FCC did not overlook the disadvantages of monopoly in the system they proposed. To guard against this undesirable feature, the Commission believed the following controls should be imposed: (1) equal representation of different carriers on the board of directors; (2) requirement for competitive bidding for the components of the system; (3) determination of the technical characteristics of the system by the FCC; and (4) the allocation of channels among the carriers by the FCC.[6] These safeguards, in combination with continuing surveillance by the Government would, in the Commission's view, make ownership by the carriers, with all its advantages, the preferred solution.

The FCC was a minority of one on the interdepartmental committee. Most participants, all at a level equivalent to that of Assistant Secretary, did not believe that government regulation would remove the dangers of AT&T domination over a purely carrier-owned corporation.[7] Although the FCC scheme would have limited AT&T representation on the board of directors, this could very well have become a legal facade hiding the underlying structure of control. Hence, the weight of opinion favored a combination of external regulation with an internal system of checks and balances. The latter would be created by extending ownership beyond the carriers to the public at large. The former would include present government regulatory powers modified and clarified to fit the circumstances surrounding the establishment of a government backed corporation.

While the NASC and the Department of Justice were preparing a legislative proposal along the above-mentioned lines, Senator Robert S. Kerr of Oklahoma announced in November, 1961 that he would present legislation in the second session of the 87th Congress favoring ownership by the carriers.[f] This political move backed up the approach of the Ad Hoc Carrier Committee publicized in a report issued on October 13.[g] To the public, it must have seemed that carrier ownership was the likely solution, supported as it was by AT&T, the FCC, and Senator Kerr, a leader in the Senate and Chairman of the Committee on Aeronautical and Space Sciences. The Welsh Committee, however, was not

---

[f]Interview, Dr. Edward Welsh, October 12, 1965.
[g]See Chapter 3.

swayed by these outside pressures. In fact, it was stimulated to submit its recommendation to the President at an early date. Thus, late in the fall, a draft bill was delivered to President Kennedy. He studied it very thoroughly and approved it with only minor changes. The plan was then submitted to the Bureau of the Budget where it received final clearance. And, on February 7, 1962, the President sent the proposal to the Hill.

## Congressional Attitudes: Hearings February to April, 1962

Within Congress, as within the Executive and the business community, opinion on the ownership and operation of communications satellites varied between the two extremes of government and carrier ownership. In the Senate, hearings on the proposed legislation were held by the Committee on Aeronautical and Space Sciences, the Commerce Committee, and the Subcommittee on Antitrust and Monopoly of the Judiciary Committee. In the House only the Interstate and Foreign Commerce Committee held hearings. The hearings of the committees generally complemented each other with respect to support of the President's proposal, albeit with modifications. There was one exception, however; the Antitrust and Monopoly Subcommittee, chaired by Senator Estes Kefauver of Tennessee, was opposed to private ownership as a violation of the antitrust laws and a gigantic giveaway of government investments in communications satellite technology.

Let us now examine these hearings to clarify Congressional attitudes on the issues involving foreign relations and the interaction between foreign and domestic policy. The results of an examination of Congressional opinions on these issues will provide data for comparing the role of Congress as an evaluator of problems falling within these two categories, to the role of the Executive branch.

In the Senate, Senators Kerr of Oklahoma and Magnuson of Washington (Chairman of the Commerce Committee) introduced the President's bill on February 7 as S.2814. On January 11, Senator Kerr had introduced his own bill, S.2650, which favored ownership by the carriers. Also in the Senate, Senator Kefauver introduced a bill, S.2890, on February 26. This proposal backed government ownership and was subscribed to by Senators Gore, Morse, Yarborough, Gruening, Burdick, and Neuberger. In the House of Representatives, a similar range of approaches to ownership were offered as bills. Congressman George P. Miller of California, Chairman of the Committee on Science and Astronautics, introduced H.R. 9696, which was identical to Senator Kerr's personal proposal. A similar bill was introduced by Congressman Olin E. Teague of Texas. However, on February 7, Congressman Miller introduced the President's proposal as H.R. 10138. And the same bill was also introduced by

Congressman Oren Harris of Arkansas, the Chairman of the Committee on Interstate and Foreign Commerce (H.R. 10115). At the other extreme were bills introduced by Representatives William Fitts Ryan of New York and Frank Kowalski of Connecticut (H.R. 9907 and H.R. 10629, respectively). These two proposals opted for government ownership. In addition, H.R. 10722 was presented by Congressman Emanuel Celler. This bill favored private ownership but with stricter controls than were in the President's bill.

The criterion for distinguishing these bills is, as indicated, their differing approaches to the ownership question. This was the most salient political issue, and its mode of resolution would have serious consequences in both the foreign and domestic political arenas. Intimately connected with the ownership issue was the question of the degree or kind of government control or regulation of the system. Three approaches to the problem of ownership and control were taken in Congress. There were those like Senator Kerr who favored private ownership with minimal government regulation. There were others like Senator Kefauver who favored government ownership, at least for the time being. And there were those like Senator John O. Pastore of Rhode Island (Chairman of the Communications Subcommittee of the Commerce Committee) who favored private ownership, with firm government control to be clearly specified in the legislation.

We have mentioned that Senator Kerr favored carrier ownership and introduced legislation to that effect, but subsequently introduced the President's proposal. This change was the result of a long series of negotiations the Senator held with Deputy Attorney General Katzenbach. Considerable pressure was brought to bear on Senator Kerr to change his attitude toward broad-based private ownership. Eventually he agreed to support the President's bill on the understanding that he could offer amendments that would modify the Government's supervisory role. In return for this concession the Government agreed to go along with the modifications, as long as they did not endanger the principle of broad-based ownership. Behind this compromise was a feeling of urgency by the President and many Democrats to increase the legislative output of the 87th Congress.

Senator Symington of Missouri was equally worried about government supervision of the proposed communications satellite corporation. His position was based on the necessity for urgency in establishing an operational system. Thus, he was opposed to the regulatory powers of the President and the Secretary of State under the President's proposal. The corporation would have a "difficult time moving forward because you constantly would have to be checking whether everybody concerned in the various government agencies approved your policy, not just one place to which you go today (the FCC)."[8] This view ranked urgency at such a premium that analysis of foreign policy issues such as aid to underdeveloped countries was nonexistent.

In the House, Representative Oren Harris, who was to be the floor leader of the President's amended bill, did not take up foreign policy issues. He maintained that a global system "holds out the promise of spectacular progress for the future—scientific, economic, social, and political."[9] But his questioning of witnesses in the hearings did not indicate what specific benefits he foresaw communications satellites as serving. Instead, the Chairman of the Interstate and Foreign Commerce Committee was concerned with three central issues:

There seem to be three basic points that are in dispute, and on which there is a great deal of controversy. One is the problem of ownership of the Corporation itself; the other one, with reference to the ownership of what is referred to as the ground facilities; and the other one, primarily the type of stock to be issued.
    There are others, of course, many others, but, to me, those are the basic ones.[10]

The Congressman's analysis of these issues was entirely from the domestic point of view, and he tended to favor the Kerr approach. In matters of foreign commerce, Representative Harris was especially concerned that the international carriers receive a good share of ownership, with the assistance of the Government. The Government should play a promotional role if not a regulatory one. This undoubtedly would help United States commerce, which was a primary responsibility of the Interstate and Foreign Commerce Committee.

A wider scope of issues was included in the analysis by those Senators and Congressmen who favored Government ownership or stricter Government controls. On the question of urgency in establishing the system, Senator Kefauver contended that this could be done just as well by the Government as by private enterprise. In fact, private enterprise might actually delay the establishment of a system in order to guard against obsolescence of its existing equipment.[11] On the question of system choice, Senator Kefauver believed that a synchronous system would represent the best choice for a worldwide system, and that AT&T's emphasis on the low-orbit system might delay this global system.[12] The Senator said:

If you get stuck with a tremendous investment by the Government and by the Corporation in a low system and if in 6 months it develops that the higher system is more feasible, and that the Soviet Union is going to use the higher system, then we would indeed be deterred in our having the better communications system.[13]

In summary, we may say that Senator Kefauver believed that control by the international carriers might delay the establishment of a system, lead to the creation of a mediocre system, and thus lead to a foreign policy defeat.[14]

The basic thrust of the opponents of private enterprise was that a violation of the antitrust laws was in the making. They were concerned with the ability of private enterprise to establish a global system, but their basic distrust of a private venture was due to a violent dislike of monopoly. In the Populist tradition, they favored government ownership of the means of communication. The Senator from Tennessee contended that "a bill authorizing several recently competing private corporations to get together and do the same thing that they are doing, constitutes an exception to the Sherman Antitrust Act."[15] Senator Russell B. Long of Louisiana asserted before the Commerce Committee that the President's bill, or its amended version, would turn control over to AT&T. Federal regulation by the FCC was bound to be ineffective, as it had been in the past. The Senator stated:

If the Government is to sell its investment in space communications, it should be done only under conditions which would guarantee complete separation and competition with the existing system.

The Government has shown the courage to do this in the transportation field, and the wisdom of that decision has proved itself. . . .

Otherwise, the Government should develop the system with a public service corporation, patterned after the various turnpike authorities or the TVA until we are in a better position to assure that the public will reap the greatest advantage of the investment.[16]

The idea of creating a Government entity similar to the TVA was the dominant motif of the opposition to the President's bill and its modified versions. The rationale was that the Government had invested billions in research in communications and missiles, and that this should not be turned over to private enterprise to operate at a profit. Rather the Government should operate it at cost for the benefit of all taxpayers. The strength of this case was undermined by the tendency of its proponents to quote figures, such as "25 billion of Government research and development," without sources to back them up.[17] Another vulnerable point of the argument was the tendency to think that if the Government controlled the proposed system, it would cost less because there would be no profit taking. Critics of this argument pointed out that if the enterprise proved unsuccessful, the taxpayers would be saved the added expense, and that if it were successful under Government ownership, it still might not be run as efficiently as under private enterprise.[18] One is left with the impression that neither side to the controversy had an impregnable position. The efficiency of a future private or public communications satellite system was a speculative matter.

When it came to the matter of evaluating the effect of satellite communications on the general goals of foreign policy, Senator Kefauver was just as vague as Senator Kerr. The Senator from Tennessee contended that "this is our great opportunity for bringing the world together, this is the greatest opportunity for an instrument of peace and understanding around the world that has come on the scene ever, I believe."[19] But he also wanted to make sure that the United States had a better communications system than the Russians.[20] How this would increase peace and understanding was left unanalyzed. Nonetheless,

Senator Kefauver and his supporters showed a greater awareness of foreign policy considerations than Senator Kerr and his backers. The new Populists, accepting the urgency for establishing the system, cautioned that haste should not obscure the longer term implications of space communications for meteorology, navigation, and space research, as well as commercial communications. Given these many functions as real possibilities, Senator Kefauver was unwilling to rigidify the legal ownership setup for fear of harnessing future developments.[21] Once established as a private venture, the system, according to Senator Kefauver would be "virtually impossible to return to Government ownership and control, no matter how appropriate such ownership may become."[22] Besides the necessity for Government involvement in launching and regulation of domestic practices, Senator Kefauver contended the Government should play the key role in these four foreign and foreign-domestic issues:

1. Supervision of changes in the internal structure of the private corporation.
2. Guarantee that opportunities are provided for foreign participation.
3. Insurance that the private corporation would provide service to areas of the world where it might be uneconomical.
4. Regulation of the ratemaking process.[23]

Points 2 and 3 indicated Senator Kefauver's concern for establishing a truly global system as compared with one based merely on profits.[h] To Senator Kefauver, business considerations and foreign policy requirements were "so intermingled that you cannot separate them."[24] He thought that the President's bill would turn over the "power of the State Department in negotiating an important international matter like this and delegate our sovereignty to a private corporation . . ."[25]

To sum up consideration of the attitudes of Congress so far, we may say that those advocating Government ownership or control lacked clarity in the presentation of their case, but were highly motivated by a wide concern for the many complex issues. Those favoring minimum Government control had neither clear arguments nor broad-based concern with the many foreign policy issues involved. They were more cogent on the financial aspects. The spotlight of both camps was mainly on domestic considerations, and the fact that a system of international communications was being established, was often lost sight of in the heat of debate over monopoly and private enterprise. Perhaps the committees carefully adhered to their jurisdictional boundaries so as not to impinge on the authority of the Foreign Relations and Foreign Affairs Committees.

Another group in Congress favored a private system but with stricter Government controls than the Kerr contingent. The main partisan of this approach was Senator Pastore of Rhode Island. As presiding chairman of the Commerce Committee in the absence of Senator Magnuson, Senator Pastore

---

[h]Although AT&T might be seen as a profit seeking corporation, it does provide service to uneconomical areas. Costs are balanced by higher charges over well-established routes.

indicated less than complete enthusiasm with the President's proposal, as it had been amended by the Aeronautical and Space Sciences Committee. He said, "We are here to perfect this bill, not to defend what is in it."[26] The kind of perfection that Senator Pastore desired, however, was not geared to foreign policy issues. An exception to this generalization, is the interest the Commerce Committee displayed in assuring the efficient allocation and management of the frequency spectrum. The reason for this deep concern, was that this problem was specifically within the jurisdiction of the Committee. The Committee and its presiding chairman were also disturbed by the ambiguity of the Government's regulatory and supervisory powers in the legislation. This ambiguity reflected on both the foreign and domestic operations of the proposed system.

Senator Pastore was especially concerned that domination by one communications carrier (AT&T) should be avoided. This would be a violation of the public interests of the taxpayers and the consumers.[27] In light of this attitude, it is not surprising that in its report, the Commerce Committee made many amendments to the bill reported out by the Committee on Aeronautical and Space Sciences. Before discussing the details of the legislative proposals and amendments, let us turn to a consideration of the attitudes of the Executive as they were expressed by various agencies and departments at the Congressional hearings.

### Attitudes of the Executive During the Congressional Hearings: February to April, 1962

During the Congressional hearings, various Executive departments and agencies and the FCC were requested to offer their opinions on the legislative proposals. We will examine the views of the Departments of State, Defense, and Justice in addition to those of NASA, NASC, the USIA, and the FCC.[i] Foreign policy attitudes in the Executive, like those in Congress, were still in the formative stage. In general, opinion was concerned more with domestic issues and issues of the penetrated system.

### *The Views of the State Department*

The Department of State took the view that an early operational system of communications satellites under United States leadership would benefit all

---

[i]Besides the testimony of these agencies, reports from other executive agencies were also requested: Civil Aeronautics Board, Department of Commerce, Department of Treasury, Federal Aviation Agency, General Accounting Office, National Science Foundation, Securities Exchange Commission, the Small Business Administration, the Civil Service Commission, and the Government of the District of Columbia.

mankind. This position can be broken down into four propositions concerning the desirability of: (1) United States leadership, (2) urgency in establishing the system, (3) the setting up of a truly global system, and (4) the benefit of all mankind to come from satellite communications. The relationship between these values was not fully spelled out.

In achieving these imprecise goals, the State Department backed the creation of a Communications Satellite Corporation under the supervision of the Government. The fact that the focal point of United States leadership would be a private corporation rather than the Government, was not considered to be a foreign policy issue by Mr. McGhee, Under Secretary for Political Affairs. "We have not taken this question up with other countries . . . we don't consider that this is relevant to the foreign relations aspect of this bill."[28] In the President's bill as proposed on February 7, Section 402 read:

The corporation shall not enter into negotiations with any international agency, foreign government, or entity without a prior notification to the Department of State, which will conduct or supervise such negotiations. All agreements and arrangements with any such agency, government, or entity shall be subject to the approval of the Department of State.

This was certainly an increase in power for the State Department. But the Department felt at this time that satellite communications were so novel that they required novel arrangements. Whereas preexisting telecommunication system were basically country-by-country systems, "what is envisaged here is a system which, at one time, covers the globe."[29] What exactly was envisioned?

We feel that the involvement of the Department is necessary to take care of the new and drastic changes envisaged by the communications satellite concept. It doesn't mean that the usual sort of negotiations, say, once a pattern has been established, which involved purely commercial or technical considerations, would be negotiated by the Department. The Department actually does not have the facilities and it would not be proper to conduct such negotiations.

But say, the first time a contact is made with another government, looking toward negotiations of an arrangement with the Corporation, the Department would enter into the picture, would set the stage, would conduct the preliminary negotiations covering general matters, perhaps, and then would leave the detailed technical operating and rate negotiations to the Corporation, subject to its general supervision.[30]

In negotiating with other governments, the State Department's jurisdiction would "encompass such matters as illegal use of the satellites, jamming, unauthorized use of radio frequencies and the location of ground stations abroad which would serve a number of nations."[31] Mr. McGhee felt that foreign governments might be unwilling to enter into a purely commercial agreement until they had agreement on these issues.

If the State Department and other Government agencies were really to have the authority and capability of exercising these responsibilities, it is no wonder that there would be so little worry about creating a corporation to serve as a focal point of our foreign policy. But, in recognition of the strong tradition of antagonism to Government interference with business, the idea of firm Government control fell through. Senator Kerr and industry representatives forced or persuaded the State Department to change its position. Rather than conduct or supervise negotiations, it became satisfied to assist and facilitate the corporation.[32] The distinction between foreign policy and business considerations was used to explain the division of powers. State was not to interfere in business. From the weight of the testimony, it is clear that State along with the new Populists, saw an inevitable overlap between the two spheres. However, under the barrage of those opposed to government interference with private enterprise, State gave in. By doing so they assured that they would offer no stumbling block to the speedy passage of the legislation.

The United States Information Agency was also vitally concerned with satellite communications. Its director, Edward R. Murrow, considered that an operable system would be "a dramatic demonstration of our concern for the peaceful uses of outer space and our willingness and indeed our eagerness to use that communications system in the tradition of free reporting in this country."[33] He was concerned, however, that if it were to be under private control, the USIA might have to pay prohibitive costs for service. He therefore advocated special rates for the Government.[34] Mr. Murrow pointed out that to use seven circuits of a system for one and a half hours a day, at current rates, would add up to $900 million a year, about $800 million more than the USIA budget for Fiscal Year 1962.

### Other Government Views

The Department of Justice was represented by the Attorney General, Robert Kennedy, in its appearance before the House Commerce Committee. Like the State Department and the USIA, he was interested in stating the foreign policy goals of the proposed system. The Attorney General said:

We are all anxious to create a global communications satellite system as rapidly as possible. We all want the United States to lead in the peaceful development of space resources. The global communications system which we envisage for the near future has a great potential for linking the world closer together and for demonstrating ways of peaceful cooperation among nations in space activities. The President attaches high importance to this program.[35]

Mr. Kennedy was especially concerned that the benefits of the system would accrue to the underdeveloped countries. The weight of his testimony indicated that he gave priority to an urgent establishment of an operational system, over enforcing a firm construction of the antitrust laws.

This was the position of Dr. Welsh of the Space Council as well. He tried to find middle ground in order that a speedy start toward establishing a working system might begin. He showed that he was an artful compromiser. On the one hand, he admitted that the carriers within the corporation might have a motive to impede the introduction of new techniques so as to guard their present facilities. On the other hand, he thought that Government might also be guilty of delay, because of the problems of bureaucratic size and red tape.[36] In handling the argument of the Kerr camp that the legislation provided for too much Government supervision, Dr. Welsh maintained that it merely spelled out existing authority of NASA and the FCC.[37]

The issue of the degree of State Department control over the foreign negotiations of the corporation was more salient, but Dr. Welsh skirted the battleground by pointing to the State Department's traditional authority in the field of international negotiations for frequency allocations. He did not discuss the Department's role in promoting peace and understanding, participation by foreign countries, and the development of a truly global system. These were the novel goals for which novel arrangements had been deemed necessary. But no member of the House Committee on Commerce asked him about the means to pursue these ends. The weight of Dr. Welsh's testimony related to domestic matters such as the effects of the internal and external controls on the corporation in its relation with the established carriers. Dr. Welsh had said that it was "of immediate importance that we get the organization, the responsibilities, and the principles of performance started on the right track, or the long-range benefits will be seriously modified.[38]

The position of the FCC at the hearings was at the opposite extreme from the Kefauver camp. The Commissioners unanimously favored ownership by the common carriers.[39] FCC reasoning had not changed since the conclusion of the work of the Welsh committee the previous fall. In the interim, however, the Commission had worked out a plan for firm regulatory control over the corporation. The powers involved mainly related to domestic policy considerations such as control over the financing of the corporation and procurement by competitive bidding. The Commission supported State's powers under Section 402, contending that they did not in any way deprive common carriers of their existing rights to engage in operating negotiations or traffic agreements with their foreign counterparts.[40] This view did not coincide with those of Senator Kerr and industry. These participants saw a threat to the existing privileges of the carriers in the original wording of Section 402. The FCC probably thought that State would interpret its authority in line with its past operations, rather than transform itself into an active regulator of the international carriers.

NASA had neither firm ideas on ownership nor government supervision of the proposed system. With its responsibilities for research, NASA was naturally more interested in technical considerations than economic and political issues. The testimony of Space Agency officials on technical matters offered a perspective which gave more concrete meaning to two problems of satellite communications development—global coverage and system choice. Dr. Dryden of NASA pointed out that "if you take a country which has no internal telephone system or radio system, little use could be made of satellites until some such system is provided."[41] This testimony indicates that satellite communications by themselves could not bring about a truly global system of communications. This fact points to the requirement for aid in the basic media before one could realistically offer aid in communications satellites. In fact, this less spectacular aid was a normal part of the Agency for International Development's (AID) budget, as well as that of its predecessors.[42]

Dr. Dryden's testimony also pointed to the fact that there was still no technical certainty as to which system would be best. Questions had yet to be answered on such matters as the life of the satellites in orbit, physical, chemical, and other characteristics of the space environment, and the size and power of ground stations.[43] It was conservatively estimated that it would take five years before an active communications satellite system could become operational. The life of satellites at the time was only three months to a year, whereas a financially satisfactory system required a life of three to five years.[44] Ideally, the synchronous system appeared to be the most suitable, but the prospects of its ever becoming operational were uncertain. In the interests of rapid development, NASA indicated that a low-orbit system would have to come first.[45]

While primarily a research organization, NASA was playing an active role in cooperating with foreign countries during the experimental stages of satellite communications development. England and France had cooperated, at their own expense, with the ECHO satellite, and NASA, with the assistance of the Department of State, was planning future experimental ventures with them as well as Germany, Brazil, and Italy in connection with the RELAY and TELSTAR projects.[46] NASA had also asked for cooperative ventures with the Soviet Union, but the Russians were less than enthusiastic. The cooperation which NASA envisaged was in the use of experimental techniques. As far as operations were concerned, NASA had no plans because these were to be carried on by a new private corporation. Under the legislation, NASA would assist this corporation with research and development information and launching, but the agency would have no foreign policy responsibilities.

The Department of Defense had no serious interest in the legislation other than that part of the bill, Section 102 (d), which preserved its right to construct a system of its own.[47] It also hoped that plans could be made to integrate the private system with the defense system in the event of a national emergency. The main thrust of the Defense Department's testimony was that the system should be established as soon as practicable. "Practicable" in this context meant to the Assistant Secretary, John H. Rubel, technically practicable rather than politically or economically feasible.[48] He was concerned that there should be no delay due to the involvement of the State Department or the United Nations.

The Department offered a different perspective from NASA's on the technical question of system choice. The Defense Department was working on Project ADVENT, a synchronous system for which it held high hopes that were to prove ill-founded. Nonetheless, at the time, Defense was more optimistic on the possibility of an early synchronous system than NASA. Defense even contended that the ADVENT system could be used for commercial purposes, but admitted that "there might be some administrative problems associated with it."[49]

## The Attitudes of Interest Groups

Most of the nongovernmental groups to express their opinions before the Congressional committees were common carriers. Representatives of AT&T, ITT, RCA, Hawaiian Telephone, and the National Telephone Cooperative Association were heard. In addition written statements were submitted by Western Union and the U.S. Independent Telephone Association. The views of the business community in general were represented by the National Association of Manufacturers and the U.S. Chamber of Commerce. Two manufacturers of satellite communications components were heard—Hughes Aircraft and General Electric. One labor union, the Communications Workers of America, an affiliate of the AFL-CIO, participated, and one liberal private group, the Americans for Democratic Action, presented its views. Of these thirteen groups, only two favored Government ownership—the Americans for Democratic Action and the National Telephone Cooperative Association. The electoral leverage of these two groups was minimal. One point to infer from this is that the genesis of the attitudes of those Congressmen favoring Government ownership was not in the political influence of outside pressure, but in the Congressman's own conceptions of the public interest.

The weight of pressure-group opinion favored private ownership. Five groups led by AT&T favored ownership restricted to the common carriers.[50] Four groups favored the President's original legislation, these being Western Union International, RCA Communications, Hughes, and General Electric. The first two companies were international telegraph carriers, and they feared AT&T domination of an enterprise restricted to common carriers. Hughes may have had similar feelings regarding the domination of AT&T's manufacturing subsidiary, Western Electric. To these companies, the remedy was not increased Government supervision, but widespread ownership.

As one would expect, interest group attention focused almost entirely on issues involving ownership and Government supervision. The only other salient issue was the question of control over ground stations. AT&T, ITT, and Hawaiian Telephone thought that ownership should be restricted to common carriers, while most other groups favoring private ownership felt that both the carriers and the new corporation should have the opportunity to own their own ground stations. Being a principal component of a communications satellite system, ground stations were much sought after. Many participants felt that control over ground stations would be more lucrative than control over the satellite itself.

The principal issue of ownership and control raised the specter of creeping socialism, a negative symbol of considerable emotive power. Carriers and manufacturers rallied to defend the prerogatives of management against those provisions in the President's bill which gave technical and administrative powers to the President and the Secretary of State.[51] The feeling was that too much Government supervision would smother the initiative of private enterprise. Government regulation of the traditional sort through the FCC was permissible (perhaps because it was spasmodic), but an increase in regulatory powers was clearly anathema. This it will be recalled was the opinion of Senators Symington and Kerr, but even the FCC, not notable for its past exercise of powers, thought that the purposes of the legislation required additional regulatory powers by the Government.

The perspectives of interest groups on concrete issues were almost entirely governed by self-interest. Consideration of the ramifications for foreign policy was practically nil. There were some notable exceptions to this pattern.

David Sarnoff, Chairman of the Board of RCA, testified that "competition . . . in international communications is a snare and a delusion."[52]

We only have competition with each other in the United States, but we do not have competition with the other end which controls the circuit. I say 'control' because they are in a position, when they make traffic contracts, to select any one of these organizations and they are in a position to determine how much traffic they will send over one circuit as against another . . . So all American companies are competing for the favor of a single correspondent in a foreign country.[53]

To rectify this situation, Mr. Sarnoff proposed the creation of an international monopoly of all record and voice communications whether sent by cable, satellite, or other means. The Government might participate in this venture whose principal aim would be to put the United States on an equal basis with foreign countries and strengthen our position in the whole field of international communications.

An awareness for the foreign policy repercussions of a private venture in space communications was shown by S.G. Lutz, Chief Scientist of Hughes Research Laboratories. He reported to Senator Kefauver that "the Latin American nations tend to favor intergovernmental or U.N. ownership and operation of a satellite communications system because many of them feel that they were exploited, in past years, by privately owned communications companies.[54] While Hughes favored a broad-based private enterprise, it was not blind to the disadvantages of this approach. In fact, this is about the only instance where a company, or at least one of its representatives, pointed to a disadvantage of its own approach.

To Congress was given the responsibility of accommodating the various views of industry, the Executive, and its own members. Hence, it is now fitting to turn to a consideration of Committee reports and floor action. The public hearings had given all interests a chance to have their views heard, if not entirely digested. The hearings allowed interested observers, such as Congressmen on other committees, to gain some perspective on the issues and the political alignments.

## Committee Reports and Floor Action

### The Committee on Aeronautical and Space Sciences

The first committee report on the President's legislation was issued by the Senate Space Committee on April 2.[55] The committee reported favorably on the bill, with amendments. Several of the amendments involved domestic questions on the composition of the board of directors of the corporation, the cost of the stock, its division between the communications carriers and the general public, and the ownership of domestic ground stations.

Let us now consider the amendments and the deletions whose substance relates to American foreign relations. The Space Committee offered two amendments to the President's original proposal. In the original proposal, the President had been given the power to "plan, develop, and supervise the execution of a national program for the establishment, as expeditiously as possible, of a commercial communication satellite system."[56] The Committee substituted language which gave the President power to "aid in the development and foster the execution of a commercial communications satellite system." (Sec. 201 (a) (1).) This amendment was not explained. Its intent was obvious, however. The President's control over the corporation would be weaker than in the original. The aim appears to have been to favor Presidential promotion of the corporation rather than Presidential regulation.

The second amendment of Senator Kerr's Committee relates to purely foreign policy considerations. Section 402 was completely revamped. Instead of having power to *conduct or supervise* foreign negotiations of the corporation, the Department of State was given the role of *advising* the corporation of foreign policy considerations. The jurisdiction of State had been seriously constricted. The intent was to draw a distinction between foreign affairs and business affairs.

In addition to these two amendments relating to foreign affairs powers of the Government, the Space Committee made certain deletions from the President's bill which touched on matters of foreign policy. In one case the Committee deleted language which would have given the President the power to determine the most constructive role for the United Nations. The Committee Report stated that this was entirely within the discretion of the President in any event.[57]

The Committee also eliminated language which would have required the FCC to *specify* technical characteristics of the operational system after consultation with appropriate executive branch agencies. Instead the FCC was to *approve* technical characteristics of the system. The report stated that the original provision was unnecessary because the FCC already consulted with other agencies as a matter of course.[58] But the Report did not mention why *approve* had been substituted for *specify*. A change in intent appears obvious. The power of specification would have given the FCC and other agencies, including the State Department, the opportunity to assure that technical choices reflected broad political goals. Instead, the Committee again indicated its agreement with those private companies which objected to government supervision over "purely" business considerations.

## House Action

In the House, Congressman Oren Harris introduced, on April 2, H.R. 11040 which was identical to S. 2814 as amended by the Senate Space Committee and reported on the same day. On April 24, 1962, the Committee on Interstate and Foreign Commerce reported favorably on H.R. 11040 with only minor modifications which did not relate to matters of substance.[59] The Committee did stress, however, the need to decide on ownership in order to present a more cohesive position at the forthcoming Extraordinary Administrative Radio Conference. The bill then went to the Rules Committee and, on the direction of this committee, was called to the floor of the House on May 2.

On the floor, Congressman William Fitts Ryan (D., N.Y.) tried to amend the proposal to allow for Government ownership, but his substitute was rejected. Opposition to the bill was also expressed by several other representatives including Congressman Emanuel Celler, the Chairman of the Judiciary Committee. In general, however, there was wide support and real enthusiasm for the legislation. Four amendments were approved on the floor, but they were refinements, not changes of substance. As a result of the ground swell of opinion favorable to the legislation, H.R. 11040 passed the House on May 3 by a vote of 354 to 9.

## Return to the Senate

Following passage in the House, the President's bill, as amended, was sent to the Senate where it was referred to the Commerce Committee. This committee had already held public hearings on the legislation in April. Hence, in acting on H.R.

11040, the Committee struck out the enacting clause and inserted in its place the body of S. 2814, as the Committee had amended the bill reported to it by the Senate Space Committee.[60] The amendments of the Commerce Committee generally strengthened the regulatory provisions and spelled out more specifically the responsibilities and authority granted. Thus, in Title I, an amendment provided that the activities of the Corporation be consistent with the Federal antitrust laws.[61] The scope of the President's power was given greater recognition when the Committee provided that he should aid in the planning and development of the system. The Space Committee, it will be recalled, had deleted language authorizing the President to plan, develop, and supervise the establishment and operation of the system. Now the Commerce Committee, under the interim leadership of Senator Pastore, reinstituted part of the original authority.

From the point of view of foreign policy, the President's powers were recognized when the word "general" was eliminated from his power to "exercise such 'general' supervision over relationships of the corporation" with foreign entities. The Committee had felt that this word suggested a restriction of the President's foreign powers.[62] Perhaps the most significant clarification of the Committee was its analysis of the role of the State Department in foreign negotiations. The report stated that:

Section 402 should be read with section 201 (a) (4) as both are concerned with the role of the corporation in relation to U.S. foreign policy. Together these sections assure that this role will be carried out in a manner which contributes to the success of that policy. Section 201 (a) (4) recognizes the President's authority to take whatever steps he deems appropriate to assure that the relationships of the corporation with foreign governments, entities, or international agencies are consistent with the foreign policy of the United States. This section reaffirms the traditional responsibility of the President, and through him of the Department of State, for conducting foreign policy. Section 402 on the other hand is concerned with the narrower problem of the corporations business negotiations with international or foreign entities.[63]

From this analysis, one can see that the Commerce Committee did not want to infringe on the foreign affairs powers of the President. Rather, the Committee wanted to draw a distinction between these powers and the customary rights of private enterprise to handle its own business affairs. We have seen that this distinction may be difficult to draw in practice. Nonetheless, it is perhaps necessary to make it in the abstract. It was now up to the President, the State Department, and other participants with competence in foreign affairs to work out the meaning of the distinction in practice.

The Commerce Committee offered several additional amendments which generally increased the powers of the FCC over matters such as the capital structure of the corporation and the regulation of competitive bidding for components of the system. The amendments stated with greater succinctness the powers of the Government, thus relating admirably to the comments of Senator Pastore that the Committee was going to perfect the legislation, not defend it.[j]

Having gone through the Commerce Committee, the bill was brought up on the floor of the Senate on June 14. There it was subjected to a bitter attack for several days from the colleagues of Senator Kefauver. The Democratic leadership withdrew the bill on June 21 in order to allow consideration of other important legislation. On July 26, the battle on the floor resumed. In the interim, an important event took place which dramatized to the public and the world at large the potential of space communications.

On July 10, the TELSTAR satellite achieved successful orbit. It relayed television, telephone, photofacsimile, and high and low speed data between the United States, Great Britain, and France. This experiment, it will be recalled, was the result of an agreement between NASA and AT&T. AT&T had negotiated with Great Britain and France to receive signals. NASA had assisted the telephone company in this endeavor as well as in providing the launch vehicles and setting the test specifications for the spacecraft. The FCC had granted AT&T a license to use frequencies. In short, this was an example of complete government-industry cooperation. To the world, the experiment appeared as an undertaking of the United States, not of the Government or of private enterprise.[64] From quarters not normally pro-American came praises for the new means of communications. Cairo said, "The launching by the United States of her satellite . . . is considered the beginning of a new era . . . and a real first step toward the use of space for peaceful purposes." From Mexico City came the message, "It is proof that science can be at the service of peace and agreement between men."[65] These tributes, which could be extended, lent weight to the argument of those who saw satellite communications as serving the cause of peace and understanding—at least on a verbal level at that moment in time.

Within the United States, the stature of AT&T zoomed. But this prestige did not extend to the idea that satellite communications should be owned by the carriers. AT&T recognized this and gave up its fight for this solution to the ownership problem. Instead it began to push actively for passage of the President's bill as amended. The rationale for this switch was probably that without this bill there would be no bill, and thus the continuation of development would take place within the Government. In addition, AT&T believed that the amendments tended to lessen the authority of the Government within the Corporation. With the carriers having the only experienced representatives on the board of directors, they would still be in a position to guard their overall interests.

Turning back to the floor fight in the Senate, we see that the cooperation between Government and AT&T in the case of TELSTAR was not sufficient evidence to convince the opposition that the pending legislation was in the best

[j]See p. 57.

interests of the United States. On July 26, a motion was made to consider the bill the pending business of the Senate. This motion itself was subjected to a six-day filibuster. On July 30, the Democratic and Republican leadership announced it was ready to file a cloture petition limiting debate on the motion for consideration. However, two days later, the leadership agreed with the opposition that the bill should be referred to the Foreign Relations Committee, but with the instruction to refer it back by August 10 as the pending business. This move was a compromise. The opposition felt that the foreign policy aspects of the bill had not been seriously analyzed.

*The Hearings and Report of the Senate*
*Foreign Relations Committee*

The Foreign Relations Committee began hearings on August 3. These gave the opposition to H.R. 11040 an additional opportunity to amend or kill the bill. However, there were only three opponents on the Committee—Senators Wayne Morse of Oregon, Albert Gore of Tennessee, and Russell B. Long of Louisiana. The other members of the Committee seemed anxious to speed the bill through, and they did not appear to devote much time to the analysis of the bill. The three opponents dominated the proceedings with their extensive inquiries, but they did not sway the predilections of their colleagues. In addition, the three Senators were met by a united front from the Executive Branch. For the first time, the Secretaries of State and Defense took part in the hearings. Also, the FCC had switched its opposition to the bill to one of firm support.[66]

Among the many issues the Foreign Relations Committee discussed was the distinction between foreign policy and business policy. The Secretary of State admitted that the two were intertwined and that in practice the Government would be faced with the problem of distinguishing between them when possible. He said:

Although the bill contains no definition of "business negotiations," this creates no difficulty. It is, after all, for the President to determine, under section 201 (a) (4) as well as under his constitutional power to conduct foreign policy, what negotiation should be conducted by the Government and what may be left to the corporation.[67]

The solution to the difficulty in practice would not lie in defining "business negotiations" and "foreign policy," but in the power of the President to decide for the Government, which is which. But the opposition was still irreconcilable. They saw in the history of Section 402 a move to delegate foreign policy to a private corporation.

The hearings ended the way they had begun—in disagreement. The Committee reported the bill favorably on August 10, over the opposition of Senators Gore, Long, and Morse.[68]

*Return to the Floor and Final Passage*

After approval by the Foreign Relations Committee, the Communications Satellite Act became the pending business of the Senate on August 10. The opposition immediately initiated a filibuster. The leadership filed a cloture petition and, for the first time since 1927, the Senate voted on August 14 to invoke cloture by a vote of 63 to 27. Following this historic procedural action, the Senate directly rejected two amendments and tabled 122, all of which had been put forward by filibuster to delay passage. Some amendments were accepted, however. On August 13, the Senate accepted an amendment by which it was made clear that the President may use Government-owned satellites rather than a commercial satellite system, if required in the national interest (Sec. 201 (a) (6) ). On August 17, the amendments of the Commerce Committee to H.R. 11040 were approved. And, culminating floor action by the Senate, the bill was passed on August 17, 1962, by a vote of 66 to 11.

The next step was approval by the House of Representatives. The floor manager, Representative Oren Harris, contended that time was running out and that further delay would damage the program. If the House offered amendments to the Senate version, the bill could then be subjected to filibuster again. Consequently the majority voted down the amendments of the small group of dissidents. The bill won approval on August 27. On August 31, the President signed it. He said the passage of the Communications Satellite Act was "a step of historic importance," and a project in which "the rigor of our competitive free enterprise system will be effectively used in a challenging new activity on the frontier of space."[69] To the outside observer, this appraisal might seem more of a ritual, chanting the language of the debate, rather than a prediction or insight into the future operations of the Communications Satellite Corporation. Let us now make our own conclusions about the Congressional stage of the policy-making process.

## Conclusions

The Communications Satellite Act is not a statement of clear policy, but a matrix of possible policies. The passage of the Act culminated a process which had been characterized by bargaining and compromise but also by polemics and ideological involvement. The focus of debate was the ownership issue. Modified private enterprise was the victorious principle. Government ownership was a lost cause from the beginning. But the dichotomy between the two was devious. It obscured the question of policy goals and their implementation. Broad parameters were set, but concrete, operational practices were left unspecified for the most part. A defender of the Act could call this flexible wording; a critic, wishy-washy language.

This situation does not mean that the policy was irrational. It is true that a premium was put on agreement rather than decision. But one could not realistically expect that clarity would characterize an act whose operation involved so many future imponderables. If the passage of the Act is viewed merely as one stage of the decision-making process, then it is possible to see rationality, clarity, and consistency occurring by stages.

The role of Congress in the decision-making process was certainly to bring greater clarity and perspective to the issues involved. If the decision had been made within the Executive without public debate, the meaning of policy, both foreign and domestic, would have been even more vague. In fact, the initiative Congress showed in amending the Act and offering alternatives is contrary evidence to the oft-repeated generalization that the Congress has declined as a legislative body compared to the Executive. It is true that the Executive proposed policy, but only because it realized that Congress would insist on a public hearing of the alternatives. Furthermore, Congress did not merely dispose. It modified, clarified, and legitimized the setting up of the Communications Satellite Corporation.

The policy process which was centered on Congress in 1962, provides some data for understanding two aspects of this book: (1) the impact of technological change on policy and the policy-making process and (2) the pattern of rationality in the foreign policy process. In 1955, John von Neumann wrote, "All experience shows that even smaller technological changes than those now in the cards profoundly transform political and social relationships."[70] While this may be true of some effects of space communications, it is not true of its effects on the policy-making process. The political and social relationships within Congress, and between Congress and the Executive, remained steadfast and unchanging. The only novelty of debate on the Communications Satellite Act was the voting of cloture to stop filibuster, and this action was obviously not the result of technological change. So, while technological innovation created the necessity for new policy, it did not alter the process by which policy was made. Technological change did relate to the competences of the various participants in the policy process, however. For instance, it became incumbent upon the State Department to have a science advisor, a position created in the early 1950s vacant before Sputnik.

As regards the second time, rationality in the foreign policy process, one can see that there was a relative lack of analysis of the consequences of the legislation on American foreign relations. Perhaps most participants thought that the foreign policy issues could be dealt with later. Why encumber an already overly complex matter now? Critics of this view might point to the fact that the ownership issue was a foreign policy decision. Most foreign governments would probably have preferred to negotiate with a system owned by the U.S. Government. Ramifications of this possible attitude were not discussed. Foreign policy consequences were irretrievably decided without a conscious decision. One answer to this line of argument might be that the ownership issue was more crucial for domestic politics than foreign politics. There had to be give and take somewhere. No piece of legislation could be perfect. Therefore, the creation of the Communications Satellite Corporation was, by and large, a commendable resolution of the issue. The more crucial foreign policy issues of program content and destination would have to be settled later.

While not wishing to take a political position on this issue, it is still proper to indicate some of the failures of the policy process to clarify aspects of American foreign policy which were not beyond the bounds of careful consideration during 1962.

One motif of the Communications Satellite Act debate was the necessity for urgency in settling the ownership issue. The widespread feeling of being pressed for time was the result of a perceived threat from the Soviet Union. It was also the result of a domestic factor, the feeling by the Democrats that they had to enhance the poor legislative record of the 87th Congress. But the new Populists did not see the necessity for settling the ownership issue in a hurry, although they saw the necessity for establishing an operational system as soon as practicable. Proponents of the ownership solution adopted never clarified why the ownership question was inherently related to the establishment of an operational system as soon as practicable. Why couldn't the Government put up an operable system now and turn it over to private enterprise later, if it was deemed desirable? Many proponents of the President's legislation would recognize that a global system was a long way off by indicating that the system would become slowly operational and then in the North Atlantic area. To say that one needs a global system as soon as possible and then say that even a regional system will be a long time in coming tends to be inconsistent. We see two interrelated ambiguities here: (1) the relationship of ownership to the goal of establishing an operational system as soon as practicable and (2) the operational significance of the goal of a truly global system.

One facet to the debate on foreign policy was the peace and understanding issue. How might someone go about presenting a case for the proposition that space communications would contribute to peace and understanding? Mr. McGhee of the State Department contended that this new means of communications would tend to break down present geographical boundaries.[71] To be more specific, this proposition might be stated thus: "Telephone, telegraph, television, and other means of communications now flow predominantly within the bounds of nation states, but space communications will reverse this trend." It is hard to defend this point of view. It confuses the means of communication with the people who are communicating. Satellite communications will increase the capability of people, groups, organizations, and computers to communicate, but it will not by itself change the identity of those communicating. Most people communicate with their friends who speak the same language and are a product of the same cultural, family, or national background. In the United States in 1962 average *daily* telephone conversations amounted to 319,838,000.[72] For the whole *year* of 1963, there were only 4,546,200 overseas calls.[73] For 1980, we have seen that the forecast was 100,000,000 overseas calls which would be only about one-third of the domestic 1963 calls per day.[k] This kind of data is true of other means of communications and other years and other countries.[74] From these "irreducible and stubborn facts" it is hard to draw optimistic conclusions about changes in the flow of communications, as is implied in the comments of those who saw space communications as a means of blurring the significance of national boundaries.

A third area of ambiguity in the foreign policy aspects of the debate was the issue of leadership for the United States in international communications. What are the various possible meanings of the term "leadership?"

1. Being first, i.e., beating the Russians
2. Whether first or not, assuring that the United States backed system is truly global and contributes to the development of the less-developed countries
3. Dominance by American industry in supplying components for any international system
4. All of these

It would seem that the third possible meaning of leadership could conflict with the second. Dominance by American industry could also conflict with the goal of foreign participation in the system. Others may not wish to cooperate where the United States dominates. As no definition was given of leadership in the policy-making process, the way was naturally left open for differing interpretations in practice, especially in light of the fact that the organizational setup was not to be hierarchically structured.

One may also see that American leadership as an aim could conflict with the goal of having one system. A truly global system would probably require some means of representing sovereign states as equals. Weighted voting in favor of the United States would probably not appeal to a great many countries.

A fourth area of ambivalence in the discussions of American foreign policy and space communications was the matter of participation in ownership and use of the system by foreign governments. One might ask which part of the system was participation to be in:

1. Research and development
2. Manufacture
3. General policy-making, e.g., over program content
4. Rate-making
5. Profits and investments
6. Ground stations, satellites, or launch vehicles

---

[k]See Chapter 3.

A list like this points to the intangible nature of the fact that agreement was reached on the policy of promoting participation by foreign countries.

A fifth area of uncertainty lay in predicting the future relationship of the new Communications Satellite Corporation and the Department of State. It became clear that State was to have authority under the President for foreign relations and the Corporation for business negotiations, but no procedure for deciding on the specific meaning of these categories was mentioned. Of course, we have seen that the modifications in Section 402 were played down by the State Department, which clarified its role by pointing to the constitutional powers of the President. But whether or not this had any legal significance to the conduct of American foreign relations, the history of Section 402 certainly had emotional significance to certain segments of the Congress and the business community. Read in the context of the debate, the modifications of Section 402 went a long way toward defining private business considerations more broadly than foreign policy powers.

Let us now turn to a consideration of developments in space communications since August, 1962. These can most fruitfully be divided into two stages, the first from August, 1962 to August, 1964 during which frequencies were allocated and arrangements made to establish an interim international system for space communications. One may date the second period from August, 1964 to August, 1971. In this stage the initial arrangements were tested and patterns were set which influenced the establishment of permanent arrangements which were negotiated between 1969 and 1971.

# 5

## The Interim Arrangements for a Global Commercial Communications Satellite System

In order to establish an operational space communications system, decisions had to be made concerning frequency allocations for the new technology, and the political and economic institutions, on an international level, best suited to promote the goals of American policy. In 1963 and 1964, there were two sets of meetings to deal with these matters, one concerning frequency allocations and the other the institutional mechanisms for providing global service. Let us now briefly describe and analyze the politics of frequency allocations in 1963, before turning to analysis of the more competitive politics leading to the establishment, in 1964, of an interim system for commercial uses of communications satellites.

### Frequency Allocations and Regulations for Space Communications

One of the principal aims of the Communications Satellite Act is to promote "efficient and economical use of the electromagnetic frequency spectrum" (Section 201 (b)). In fact, achievment of this goal had received careful and continuous attention from Government agencies for a long time prior to 1962. More time and effort was spent to attain this objective than all of the other foreign policy objectives combined through 1963.[a] It will be recalled that the 1959 Administrative Radio Conference had allocated 13 narrow bands for experimental use by space services.[b] The Conference recognized that this allocation would serve only for a short time, and, hence, it called for an Extraordinary Administrative Radio Conference (EARC) to be held before the Ordinary Radio Conference in 1965.

Preparations for this conference were undertaken by various technical committees under the auspices of the ITU and several member nations of the ITU. In the United States preparations were under the dual control of the FCC and IRAC, the FCC being the decision maker for nongovernmental services, and IRAC, an executive committee where negotiations and logrolling between different government agencies took place.

In contrast to many other countries, United States proposals not only had to be geared towards an international audience but also had to be acceptable to diverse governmental and nongovernmental claimants which were not hierarchi-

---

[a]This is a personal estimate based on my research.

[b]See Chapter 3.

cally organized. The complexity of the internal decision-making process was not a drawback, however, for in 1962 the United States was able to present its preliminary views abroad before other countries had made their decisions. It was a definite advantage to control the agenda of the preconference discussions.

The Extraordinary Conference convened on October 7, 1963 in Geneva, Switzerland. The United States was represented by forty-three delegates chosen from the Executive, Congress, industry, and the academic community.[1] The composition of the delegation was itself witness to the emphasis within the United States on the importance of frequency allocations for space services. The Chairman of the American delegation was Mr. Joseph H. McConnell, President of Reynolds Metals.

At the time of the EARC, there were 122 members and one associate member of the ITU. However, only 70 countries participated in the work of the conference. The question immediately arises as to the reason for this meager turnout. Many countries from Latin America and Africa south of the Sahara did not attend. Perhaps as the Secretary General of the U.N. had written in his report, these nations were too concerned with their needs for traditional forms of communication as well as their grave problems of poverty, disease, and hunger. It is significant that the director of communications systems for NASA had written earlier in the year:

Communications satellites will aid the less developed countries only to the extent that those areas possess or acquire telecommunication nets, educational plants and socioeconomic systems capable of distributing and gainfully using the telecommunications made available by communications satellites. All nations will not benefit equally from participation in a world-wide communications satellite system; indeed, some nations perhaps should not participate at all. Clearly, only a small number of countries should have satellite ground stations . . . [2]

Perhaps many countries believed this assessment. That would explain their absence from the Extraordinary Conference.

Most of the work of the Conference took place in committees and working groups. The three major committees were the Technical, Allocation, and Regulation committees.[3] Allocations and regulations concerning other services besides communications satellites were discussed and resolved, but here we shall restrict ourselves to a consideration of the problems and decisions relating to the communications-satellite service. There were essentially four problems involving space communications: (1) direct broadcasting from satellites, (2) frequency allocations for satellite communications, (3) the role of the ITU as a coordinator, and (4) the provisional or nonprovisional status of the allocations. We shall deal with these issues in order.

*Direct Broadcasting from Satellites*

The first problem came up in Working Group 4B where France and the United States offered contradictory proposals. The French proposal would have prohibited the television or radio broadcasts from any space object. The American proposal would have permitted experimental broadcasting in technically suited bands then allocated to the broadcast service, pending further studies by the CCIR.[4]

In the course of the consideration of these proposals, the Soviet Union gave its support to the United States position. The French-American difference was resolved by a compromise according to the report of Mr. McConnell, but the decision was to allow experimental broadcasting "without calling special attention to this fact." Presumably this meant that the decision was less authoritative than it would have been otherwise.

The controversy over broadcasting shed light on the inadequacy of the allocation categories in light of the development of space communications. Did the broadcast category include all broadcasts or just those from the earth? A partisan of space broadcasts could argue that the old broadcast allocations logically covered this new area unless the ITU took a conscious decision to the contrary.

*Allocations*

The second major decision was to allocate 2,675 megacycles of the radio spectrum to the communication-satellite service. The proposals of Canada, France, and the United States were logically identical and added up to a total of 2,725 mc/s. The British proposals also corresponded except that they included an additional 500 mc/s. The Soviet Union, however, wish to allocate only 1,600 mc/s of which only 800 mc/s corresponded with the United States proposal. In addition, the Soviet proposal did not include any provision for exclusive use whereas the other proposals allowed 100 mc/s. A complete impasse developed in Working Group 5A as the conference came to the end of its third week.

To determine whether any compromise could be reached, the heads of the French, British, Soviet, and American delegations held exploratory talks. Fortunately, a compromise was reached. In the words of Mr. McConnell, it had two principal aspects:

(a) the willingness of the U.S.S.R. to agree to more spectrum space for the communication-satellite program than they felt to be necessary during the life of the present agreement, and the corresponding willingness of Western delegates to reduce the amount of readily useful spectrum space in their proposals from 2725 mc/s or more to 2000 mc/s; and

(b) agreement that certain bands containing high-powered operations be exempted from the technical criteria and coordination procedures prescribed under the regulations for spectrum space shared by the communication-satellite service with existing services.[5]

The compromise meant, however, that 800 mc/s of the allocated spectrum would not be useful to the United States due to interference with established services. But the 2,800 mc/s allocated[6] would be adequate to handle something like 8,000 to 9,000 2-way voice channels and perhaps four duplex television circuits. Considering the fact there were only about 550 overseas telephone circuits in 1961 and that the projected need for 1980 was 8,000, the 1963 allocations of the EARC were quite adequate.[c]

One rationale for allocating such an extensive amount of spectrum space was to provide space for defense purposes of the United States, the Soviet Union, and other countries with space communications potential. The military requires considerable amounts of spectrum to prevent attempts at jamming and to communicate with low power receivers.

*Regulatory Role of the ITU*

The third problem concerned the coordinating role of the ITU. France proposed that it be made mandatory to notify the IFRB of all assignments to terrestrial stations to be operated in bands shared with the space vehicles. The United States, United Kingdom, and Canada successfully opposed this proposal on the grounds that the administrative burden of implementing it would outweigh the advantages.

The French also proposed that all space services be coordinated in advance with members of the ITU. This was also opposed by the United States on the grounds that it was unnecessary. The reason for the American position was not clarified in the McConnell report. The French argued coordination was necessary in order to obtain parity for the treatment of space and terrestrial services. The disagreement was ultimately settled by an extensive revision of Article 9 of the 1959 Geneva Radio Regulations. This assured equitable treatment for the space services and the fixed and mobile services in the applicable shared bands.

The United States was consistent in its opposition to greater coordination. It objected to a Soviet proposal which would have required all ITU members to coordinate their proposed space system with all other existing or planned space systems. The United States felt that this plan could permit one country to block another's space plans. A compromise was again reached whereby informal coordination through the IFRB was decided upon, in addition to allowing one nation to submit comments to another's plans.

cSee Chapter 3.

*The Provisional or Nonprovisional*
*Status of the Allocations*

The fourth major issue was whether to allocate the agreed frequencies on a definitive basis. The USSR contended that allocations and uses should be interim in nature. This position was supported by Israel in a document which stated that space communications is "both the privilege and the exclusive possibility of the great countries only," that the duty of the "Space Conference is to abandon or at least modify the present practice of first come first served," and that "some form of a Space Communication Administration may be set up entrusted with the responsibility for insuring the global interests. . . . of all member states of the Union."[7] A definitive allocation would make it easy for the great countries to establish working systems, but, if and when other nations wished to set up systems, they would find that the frequency allocations had already been used up.

This view was also supported by the expert opinion of the International Frequency Registration Board. The United States delegation protested the Board's authority to submit recommendations of this kind and the proposal was formally withdrawn. But the Soviet proposal still remained. Fortunately for the American position, the proposal to consider the decisions of the Conference as interim was defeated by the Regulations Committee by an informal vote of 18 to 4.[8]

Joseph Charyk, the President of the Communications Satellite Corporation, thought that this decision was essential to a successful commercial venture:

. . . there is now a basis for, if you will, an investment based on some assurance that the whole thing isn't going to be upset by another look at the matter in a few years without any positive decisions having been taken . . .[9]

In conclusion we may say that the results of the Extraordinary Conference were very favorable to the United States. The Conference adopted the majority of the United States proposals in substance. To Mr. McConnell, this result was a tribute to the value of the thorough pre-Conference international coordination carried on by IRAC, the FCC, and the Department of State.[10] The domestic policy-making machinery had brilliantly met the challenge of international bargaining for frequency allocations.

It may also be pointed out that the international process on frequency-allocations decisions was legislative in character, as was the domestic process. This fact has been recognized by Donald R. MacQuivey, who wrote in 1962 that "the preparation and adoption of the provisions of international agreements (on radio frequencies) may be characterized as international legislative action because they are somewhat analogous to the preparation and adoption of domestic legislation."[11] The reason why national antagonisms and jealousies did not overwhelm the proceedings was the existence of a common need for increased communications throughout the world. All participants recognized this need and were willing to coordinate their activities accordingly. Coordination was not a threat to national security or prestige.

Samuel P. Huntington defines legislative decision-making in relation to three factors: (1) participants relatively equal in power, (2) disagreements about goals, and (3) the existence of numerous alternatives.[12] All these factors would appear to have characterized the negotiations at Geneva in 1963.

Let us now turn to a consideration of the negotiations for setting up the economic and political structure for a commercial system of space communications. Here, the climate of opinion, both within and outside the United States, was not as relaxed or consistently characterized by harmony and good will.

### The Establishment of INTELSAT

In the Communications Satellite Act, foreign participation "in the establishment and use" of a communications satellite system is provided for, and the Corporation is authorized to operate the system "itself or in conjunction with foreign governments or business entities."[13] What form would participation take? The statutory language is open-ended. Partnership could be in the use of the system, the ownership, or the research and development. Which particular forms would evolve was not mandated by the Act, yet it is clear that operational relationships would have serious consequences on domestic and foreign policy. The substance and the form of United States foreign policy goals to promote leadership, international peace and understanding, and the rapid establishment of a global system were involved.

In order to cooperate with other countries, it was first necessary for the United States to develop a cohesive approach towards specific problems. This involved a gradual working out of the relationship between the Government and Comsat. But early consultations and briefings on the international level occurred at the same time as domestic attitudes crystallized. While it is sometimes fashionable for, analytic purposes, to separate domestic and international discussions, in this case the separation would not be legitimate, because foreign policy consultations affected the bargaining position of the United States as it evolved from the spring of 1963 to the summer of 1964.

Let us now discuss the domestic developments whose consequences affected the formation of United States policy towards international arrangements for communications satellites. Attention will be focused on the establishment of the corporation and its relations with the State Department and other agencies responsible for foreign policy considerations.

## Relations with the Executive Branch [d]

The first step toward establishment of an operational communications satellite system after the passage of the Communications Satellite Act was to form a private corporation under District of Columbia Law.[14] It was the duty of the President by and with the advice and consent of the Senate to appoint incorporators as the initial board of directors.[15] President Kennedy appointed thirteen incorporators on October 15, 1962. The collective experience of these men in communications, international relations, and the law assured the Corporation of a wide expertise and competence during its gestation period.[16] The incorporators were political appointees, but President Kennedy combined politics with expertise. The incorporators chose Philip L. Graham, the publisher of the Washington Post, as their first chairman on October 22, 1962.

Mr. Graham's relations with the Executive were short lived as he resigned from the Corporation in January for reasons of health. During his three month tenure, however, he strenuously objected to what he saw as the interference of the State Department in the international aspects of Comsat's plans.[e] There was a difference of opinion on the decisiveness of the role State should play in establishing an operational system. State had been briefing European telecommunications administrations on American policy, and Mr. Graham thought this was Comsat's prerogative.

This attitude persisted with Graham's successor, Leo D. Welch, who had become Chairman and Chief Executive Officer in February 1963.[17] Mr. Welch had formerly been Chairman of the Board of Standard Oil of New Jersey and he was dedicated to the proposition that private initiative by the Corporation should determine policy. Welch did not see the State Department as an ally but as an obstacle. This was also true of his approach to the FCC. He charged the commission with the "invasion of the managerial functions of the corporation" in connection with the FCC's directions on how the corporation should disburse its funds.[18]

On the other hand, Comsat's relationship with NASA were always cordial. During its incorporation period, when its financial resources consisted only of borrowed funds,[19] the Corporation depended largely for its research and development upon NASA programs. Close and constant cooperation was thus a

---

[d]In this section, I draw heavily on interviews I have had with:

1. William Gilbert Carter, Special Assistant for Space Communications in the Bureau of Economic Affairs, State Department, August, 1962–August, 1964.
2. M.V. Mrozinski, Staff, NASC.
3. Daniel Fulmer, Staff of the House Military Operations Subcommittee, 1964-1966.
4. Edwin J. Istvan, Director of International Development, Comsat, 1964-1970.

[e]Interview, William Gilbert Carter, March 18, 1965.

prime necessity for economic survival. This was a symbiotic relationship as NASA had the responsibility under the Communications Satellite Act to cooperate with the Corporation in research and development.

On the whole, however, the relations between the Government and Consat were somewhat less than ideal. This was not the result of personality conflicts but of opposing views based on differences of opinion. While the Government was looking forward to close cooperation and firm regulation of the Corporation, Mr. Welch saw Comsat as a traditional private enterprise. But Government officials saw Comsat as an entirely different entity. Comsat was more a public utility, even a creature of the Government. Under the Act, it was to serve public as well as private purposes. The Corporation was to be regulated not only by the FCC as all carriers are, but by the Director of Telecommunications Management and the President in the exercise of his powers over foreign affairs.

The initial estrangement between the State Department and Comsat was especially crucial for foreign policy and foreign relations. Comsat was under no firm statutory responsibility to consult with State. Thus the leaders of the corporation could choose to follow their own individual preferences. The ability of State to persuade Mr. Welch of its prerogatives was also hampered by the State Department's weak power position in domestic politics. To overcome this disability, State tried to convince Comsat that State's views were also the Europeans' views, and that Comsat was not only a domestic corporation but the chosen instrument of American foreign policy.

While positions between the Corporation and the Government were not polarized, there were four issues whose resolution involved differences of opinion:

1. **The ownership issue**. Some participants thought the Corporation should own the entire system, ground stations as well as satellites, while others thought the system should be divided into its constituent parts, ground stations being owned by the nations in which they were located and the satellites being owned jointly by the participants in the international venture.

2. **The form of an international agreement**. Some participants thought that Comsat should negotiate on a bilateral basis with a series of countries to establish the system, while other thought the system should be organized on a multilateral basis with two separate agreements, one on an intergovernmental level and the other on a commercial basis.

3. **The management role of Comsat**. Everyone agreed that Comsat should be the manager of the international system, but there were differences of opinion relating to the degree of control between what we might call the advocates of 100 percent control and the partisans of 80 percent control.

**4. The role of the Government in the negotiation of the agreements to establish the international system.** Some corporation officials thought that Comsat should be the principal negotiator while some Government officials wanted to place prime responsibility with the Department of State.[f]

What was the process by which these issues were resolved? In June 1963, Comsat and the State Department agreed to get together to work out principles for international participation.[g] In the same month President Kennedy established an Ad Hoc Communications Satellite Group under the joint chairmanship of Nicholas Katzenbach and Dr. Jerome B. Weisner. This committee exercised the responsibilities of the Director of Telecommunications Management due to a vacancy in that office.[20] A subcommittee, composed of representatives of State, Justice, the FCC, the Space Council, and the office of the Director of Telecommunications Management, kept close tabs on Comsat's draft principles for international cooperation.

At first the ideas of the State Department did not coincide with those of Comsat on the problem of foreign participation. Mr. Welch wished to establish a Comsat-owned system by a series of bilateral intergovernmental agreements. Comsat would be the primary negotiator. This conception of international cooperation clashed with what State viewed as politically feasible and desirable. To State it seemed that some form of joint ownership would be necessary; if other nations were to have no control over the system, why should they cooperate with it?

State also thought that it would be necessary to conclude an intergovernmental agreement as well as a commercial transaction. As far back as the fall of 1962, William Carter of State's Bureau of Economic Affairs considered that negotiations on two levels would have to be undertaken in order to expedite the establishment of an international system. On the one level, there would be discussions between governments looking to the conclusion of an agreement on the principles of international cooperation in space communications. On the other hand, there would be technical and business agreements between the telecommunications administrations of the various countries. These would be analogous to the traditional agreements for the management of cables and radio. The necessity for the two-level approach was based on the assumption that other nations would be unwilling to join a venture whose political and economic consequences were completely removed from their influence. The validity of this assumption was based on the experience that Carter had had in discussing space communications with foreign countries in 1962 and 1963.

The chairman and the president of Comsat were not advocates of the same

---

[f]The conceptualization of these positions presents a clarity of roles which only developed in the summer of 1963. In the spring, positions were still imprecise.

[g]Interview, William Gilbert Carter, March 18, 1965.

line of reasoning even though they too had engaged in international consultations. In May and June they had held meetings with representatives of Western European nations, Canada, and Japan.[21] While they had not reached any definite conclusions, they favored a Comsat-owned, Comsat-negotiated venture. But within the corporation, Edwin J. Istvan, who had come to Comsat from the Defense Department's Office of Space Systems, had drafted a set of principles for international participation which came nearer to the views of State. This preliminary draft was based on three ideas:

1. That partnership in the system could take one of the following forms: space segment ownership, communications terminal segment ownership, or mutual aid participation to assist developing nations. Space segment ownership in the satellites, tracking and launching vehicles would be allocated in relation to a country's present share of world communications traffic.
2. That the ownership in the system should be established by an agreement between the Corporation and the participating foreign entity. The U.S. Government would not play a prime role, but there would be joint ownership.
3. That Comsat would be the manager of the system to the extent of controlling its own percentage of ownership and controlling contract awards.[h]

Mr. Welch did not accept these ideas. He maintained that the best system would be centrally owned, managed, and technically integrated. But the Ad Hoc Communications Satellite Group thought differently.

In the discussions that the Corporation had with this group, it became evident that the Government favored an approach based along the general lines of the Istvan draft. Istvan had played a crucial role in pointing to the possibility of separating space segment ownership from station ownership. The Government accepted this idea but not the second or third. It was contended that the Government had developed the capability at its own expense and therefore should have a central role in determining how it was to be organized on an international as well as a domestic basis. The State Department felt that other nations would not be willing to establish an international agreement for communications satellites along the same lines of the cable arrangements. Negotiating between governments would be required to lessen fears that communications satellite technology would disrupt existing investments before they had been amortized.

The Ad Hoc Group objected to the third idea not so much for its substance as its form. It was asserted that other nations would be more likely to join a system if the management role of Comsat were played down. The State Department

[h]This information is paraphrased from material in the files of Comsat.

thought the management role of Comsat should be worked out in negotiations rather than proclaimed beforehand. As the United States was the only country with developed potential, it would be natural for other nations to agree to Comsat's primary position.

The State Department was especially concerned to present a flexible and attractive proposal to other nations so as to promote the establishment of a single global system rather than the establishment of competing systems. Writing in May, 1963, the Deputy Assistant Secretary for International Organization Affairs stated that "economic, technical, and political considerations all point to the desirability of a single system:

From the *economic* point of view a single system would avoid wasteful duplication of expensive satellite and ground facilities.

From the *political* point of view a single system would enhance the possibility of fruitful exchange of communication between all countries and would avoid destructive competition to tie different countries into the communication systems of political blocs.

From the *technical* point of view a single system would facilitate technical compatibility between satellites and ground terminals, would assure the best use of scarce frequency spectrum, and would promote operational efficiency and flexibility in routing.[22]

In light of these advantages, and considering its opinions on foreign attitudes, it is no surprise that the State Department objected to the ideas of Mr. Welch and wished to modify those of the preliminary draft drawn up by Mr. Istvan. It was of the utmost importance to project an image of partnership rather than of paternalism. Otherwise, United States policy would lend weight to the Communist charges of exploitation by capitalist monopolies and, furthermore, could encourage the establishment of competing systems.

An integral part of the Government's position for a more flexible negotiating position was its contention that Comsat should undertake its discussions with foreign telecommunications administrations in light of consultations with interested Government agencies. The implications of various decisions were too far reaching to be considered merely business responsibilities. For instance, the Department of State thought that system choice "will have important consequences on coverage and burden-sharing, the availability of satellites for communication between and within countries, and the cost of ground installations which nations will have to build."[23] In addition, State asserted the following:

The establishment of global satellite communications will involve international discussion of other questions: the handling of research, development, manufacture, and launch; participation in ownership of ground terminals and satellites; allocation of satellite channels between use and users; means of determining the number and location of ground terminals; technical standardization; ratemaking; assistance to less developed countries; possible public uses of the system by government information agencies of the U.N.; and means of facilitating the exchange of programs between nations.[24]

Decisions on all these problems would require the closest possible cooperation between Comsat and the Government, according to the State Department. In many cases, the Government's role would be regulatory in addition to being promotional. But the difficulties in evolving a working statement of principles continued into the fall of 1963.

During the summer, Comsat discussed its draft principles of foreign participation with representatives of the Departments of State, Justice, and Defense, NASA, the FCC, and the President. As a result of these discussions the statement of principles by the Corporation was slightly revised, but not in substance. Mr. Welch still held to the idea that Comsat could be the owner of the entire system. The principles of the Corporation were also discussed with the common carriers and interested committees in Congress. But in the course of this filtering process, there was still no significant abridgment. The Chairman of the Senate Foreign Relations Committee, Senator J. William Fulbright, expressed dismay that Government participation in the preparation of the principles was not more apparent, but, in reply, Mr. Welch expressed the opinion that it was.[i] Other committees, however, were not opposed to Comsat's position. Thus, as the time came near to start actual bargaining with other countries, the United States negotiating position was not cohesive.

It is somewhat surprising that Mr. Welch's views had not moved more toward center in light of European attitudes. After a June meeting of potential partners in London, the British Government had sent the Department of State an aide memoire in which the British related American policy as they understood it. This brief stated that the United States favored shared ownership and pluralistic decision-making. The statement was approved by State and cleared through Mr. Charyk in Comsat before being sent back to the British.[j] Yet Mr. Welch was still arguing for 100 percent Comsat control. As late as September, he (and the carriers) objected to an Istvan draft based on the idea of two agreements—one governmental, and the other between communications entities.[k] It is true that on October 4, 1963, the Government, Congress, and the carriers cleared a Comsat draft on negotiating principles, but clearance by the Government did not mean approval.[l]

At the Extraordinary Administrative Radio Conference in Geneva,[m] an informal United States policy group had been hard at work trying to convince

[i]Letters to this effect may be found in Comsat files.
[j]Interview, William Gilbert Carter, July 28, 1966.
[k]Interview, Edwin J. Istvan, July 8, 1966.
[l]Interview, William Gilbert Carter, July 28, 1966.
[m]See pp. 75-80.

member countries of the ITU that the proposed system was not an American monopoly in disguise. Later the same month, a meeting with the European nations produced the harmony in the United States position that was lacking as a result of purely domestic discussions. An American negotiating team met with a united European group which had been formed to negotiate with the United States. This group was known as the European Conference on Satellite Communications (ECSC or, in French, CETS). At this meeting the Europeans made it known that they were unwilling to negotiate with Comsat on a bilateral basis. This blunt fact of necessity changed Mr. Welch's outlook on the character of an international system.[n]

Comsat now officially agreed to principles which envisioned joint ownership and negotiations conducted by a joint American team representing equally the Government and the Corporation. Henceforth Comsat worked more closely with the State Department on the international level. It was agreed that a joint negotiating team headed by Mr. Welch and Abram Chayes, the State Department's Legal Advisor, would represent the United States in forthcoming discussions with the European and other nations interested in establishing a global communications satellite system. The Corporation and the Government also agreed on an explicit formulation of principles to serve as the United States negotiating position. The stage was thus set for a harmonious approach to establishing a global system.

As a possible reinforcement to close relations between Comsat and the Government, several important Government officials became officers of the Corporation. In January 1964, John A. Johnson, who had been General Counsel for NASA, became Vice President in charge of the international affairs of Comsat. Lewis Meyer, who had been a Deputy for Financial Analysis in the Air Force, became the Finance Coordinator. Richard Colino, who had had positions with the State Department and the FCC, became an assistant to Mr. Johnson. In addition, Louis B. Early and Siegfried H. Reiger, both of whom had been with the RAND Corporation, joined Comsat to serve as Chief of Economic Analysis and Manager of Systems Analysis respectively. While one might have expected the Government to play an active role in promoting staffing of this kind, this was not the case. Former Government personnel came to the Corporation because it was anxious to find a nucleus of people with experience in satellite communications. Such a nucleus could only be found in the Government and among certain selected carriers and manufacturers such as AT&T and RCA.

*Congressional Oversight*

The first Congressional consideration of communications satellite developments, after the passage of the Act, came in March 1963, when the Senate Commerce

[n]Interview, William Gilbert Carter, March 18, 1965.

and Aeronautical and Space Sciences Committees held hearings on the President's nominations of incorporators.[25] Both hearings were a forum for expressing dissatisfaction with the affairs of the Corporation. The incorporators were lauded for their credentials, but criticized for their activities or lack thereof. Concern and/or criticism came from Senators who had been anxious to see the speedy establishment of an operational system under the auspices of private enterprise.[26] But, while this was the hope of the majority of Congress, according to several Senators actual practice seemed to be Government subsidy to a private corporation.

Senator Warren Magnuson, the Chairman of the Commerce Committee, allowed for NASA assistance to Comsat but contended:

The participation of the Government with respect to research shall not be of the size and of the quality to defeat the very motive that gave origin to this private corporation. We contended at the time that we formed this corporation that what we were trying to do was release the taxpayer of this responsibility.[27]

The stimulus for this comment was the NASA request for $50 million to carry on research and development in communications satellites. As the Space Agency was to have no operational program of its own, this money seemed destined to subsidize Comsat. But the Senators had no intention to let this happen. "Quite obviously," Senator Clinton Anderson said, "the Government will not spend $50 million for the benefit of a private corporation."[28]

Nonetheless, the fact was that the Corporation did not have the funds to carry on extensive research and development. Financial solvency would have to await the results of the projected stock issue. There was a disturbing paradox here, however. The incorporators of Comsat recognized that:

the American people cannot be asked to support an enterprise of prime importance to the Nation and the world without sound knowledge of the public policy, technology, and economics that will govern it; of the people who will be responsible for its operations; of the relations with governments and industry, at home and abroad, which will determine its validity; of the services it aspires to provide; of the nature and dimensions of the organization it plans to establish.[29]

But all these factors were imponderables in early 1963. Thus, there could be no stock offering and no capital of any great magnitude for the Corporation to invest in research and development. Hence, it was incumbent upon NASA to continue to carry the responsibility for research and development in space communications.[o]

The Senators were not so willing to go along with what was obviously in the private interest of the Corporation. They were concerned with the nature of

---

[o]NASA and Defense also had the duty to carry on R&D for satellite communications to meet unique government requirements.

Comsat's relation with NASA. The Space Agency had clear authority under Section 201 (b) (2) of the Act to "cooperate with the corporation in research and development to the extent deemed appropriate by the Administration in the public interest." The Corporation as a private enterprise had a clear duty to its owners to run at a profit. If it could receive technological assistance from NASA in the public interest, it would be only natural to accept. But here we see a built-in conflict of interest relating to the role of Comsat as a private enterprise and its role as the chosen instrument of American foreign policy. In the one instance NASA assistance was cooperation in the public interest; in the other, it was secret collaboration. Although Government assistance to private enterprise through placing its knowledge in the public domain is a normal practice, in the case of Comsat, such assistance would be given to a monopoly rather than distributed to competitive companies.

The Senators were quite naturally worried about the possibility that NASA might be subsidizing the Corporation, but they recognized that this possibility was probably the inevitable result of the split personality of the Communications Satellite Act itself. Senator Philip A. Hart of Michigan made the point that Comsat "from the opening day of business was faced with a tough problem because Congress really had built in a conflict of interest in the act we ought to face up to."[30] Senator Pastore said:

It is up to the Congress to set the lines and to set the guide rules with reference to who is to do what so that the public interest will be protected, so that the tax-payer will be spared the responsibility of conducting the investigation and the research which will inure to the benefit of people who will collect dividends.[31]

Perhaps the most distressed voice of concern was Senator Symington's. It will be recalled that he had supported the private enterprise philosophy wholeheartedly.P He had argued against Government supervision of Comsat, and now it appeared that NASA would have to support Comsat. Senator Symington was shocked that the Corporation had no plans to pay or not to pay NASA for its work, especially as it was a requirement of Section 201 (b) (3) and (5) that the Corporation reimburse NASA for assistance.

I have been in at the birth, and at the death, of many corporations in the past, and have never run into one like this before. Your testimony (referring to Mr. Bruce Sundlun, an incorporator), does not satisfy me with respect to my concept of what kind of corporation I was voting for last fall.[32]

Senator Symington had voted for private enterprise, but the Act really contained what Senator Hart called "a built in conflict of interest."

In light of the Congressional dialogue on the subsidy issue, it is no wonder

PSee Chapter 4.

that later in the year Congress reduced the $50 million requested by NASA for its communications satellite program to $42,175,000. In addition, the authorization contained the following language of intent:

Provided, however, that no part of any funds authorized to be appropriated by this act may be obligated or expended for the furnishing of any scientific or technological services for the exclusive benefit of any person providing satellite communications services other than an agency of the United States Government, except at the request of such person and on a reimbursable basis.[33]

The subsidy issue is also interesting from the point of view of the intertwining of domestic and foreign policy because it was linked to the sense of urgency surrounding the passage of the Communications Satellite Act. Supporters and opponents of the Act both expressed the need for speedy establishment of an operational system. The most pressing rationale for speed was the necessity to establish leadership over the Soviet Union. Behind the scenes we have seen that there was another reason—the awareness of the Democrats that the 87th Congress had a very poor legislative record. In retrospect, we can see that the filibusters had a point in contending that a private enterprise might slow up development of an operational system. Only a close Government relation with Comsat could assure speedy development. And, in the initial incorporation period, the burden was carried almost entirely by the Government.[34]

From the foreign policy vantage point, the need for haste became somewhat questionable as time went on. It will be recalled that the NASA Deputy Administrator, Dr. Hugh Dryden, had said in 1962 that it would be about five years before an operational system would be aloft.[q] The 1962 sense of urgency in deciding on an organizational alternative for an outcome in 1967 seems somewhat misplaced. But 1962 was nearer Sputnik I. What we see, then, is a psychological sense of urgency resulting from the exigencies of domestic politics combined with the after image of a past foreign policy failure. This sense of urgency in response to an imminent threat of defeat was based on an unwarranted assumption. Certainly there was a need for long-term planning in order to meet the growing demand for communications circuits and to meet the threat that the Soviets might develop an alternative system. But these reasons do not explain the atmosphere surrounding Congressional debate on ownership in 1962. The atmosphere of this debate was determined by a combination of internal and foreign expectations whose relative importance did not point to a dispassionate assessment of future contingencies.

Given this assessment, it is natural that, with a breath of fresh air and a sufficient distance from the event, many Senators were appalled at the statute they had helped make law. There was a questioning of the exigencies of complex domestic policies and relationships. It was recognized that careful oversight by

---

[q]See Chapter 4.

Congress, thoughtful assistance by the Government, and public-spirited support by Comsat would be necessary to establish operational arrangements in the national interest. It was recognized that Comsat was not a private enterprise but a sort of quasigovernmental agency or, as Senator Cannon of Nevada said, a "semipublic or semiprivate" corporation.[35] Now what was urgently needed was cooperative management by the Government, Comsat, and the common carriers.

There was concern in Congress that this need for accommodation would not be met. This problem was the other side of the coin of the subsidy issue. Whereas many felt that the relations between NASA and the Corporation might be too close, they felt that the relations between Comsat and the FCC and the State Department might be too distant. The expression of this concern by Congress helped to bring closer cooperation among Comsat, the State Department and the international carriers.

Let us now turn to a consideration of the international discussions and negotiations leading to the establishment of the Interim Arrangements for a Global Commercial Communications Satellite System. It should be recalled that they occurred both before and after the United States was able to reconcile the positions and policies of its domestic constituents.

*Foreign Negotiations for*
*International Arrangements*

The Space Age put the smaller industrialized nations to a great disadvantage as compared to the two super powers. America's allies in Western Europe and elsewhere could only develop their own capabilities in space through a symbiotic relationship with NASA. Cooperative experiments were most often arranged on a bilateral basis, but, early in the Space Age, the Western Europeans saw the advantage of building multilateral organizations to pool their resources and thus develop a technological base independent of the United States.

At the intergovernmental level, discussions leading to the establishment of the European Space Research Organization (ESRO) were begun in 1960.[36] A pact signed by representatives of Belgium, Britain, Denmark, France, the German Federal Republic, Italy, the Netherlands, Spain, Sweden, and Switzerland entered into force on March 20, 1964. The purpose of ESRO is "to provide for, and to promote, collaboration among European states in space research and technology exclusively for peaceful purposes." When one notices that the composition of the organization includes neutrals, it is not surprising that the proviso "exclusively for peaceful purposes" has more than rhetorical importance. There are no military projects. ESRO restricts its activities to those of scientific importance rather than extending itself in the race for prestige, security, and wealth.

A second European multilateral organization for space purposes is the

European Space Vehicle Launcher Development Organization (ELDO). This organization was conceived as a result of Anglo-French discussions early in 1961, and the convention establishing it entered into force on February 29, 1964. ELDO's composition is as follows: Australia, Belgium, France, the German Federal Republic, Italy, the Netherlands, and the United Kingdom. The purposes of ELDO have more scope than those of ESRO. ELDO hopes to develop operational programs in meteorology, navigation, and telecommunications as well as generating knowledge of pure science.

A nongovernmental, multilateral, nonprofit organization of considerable importance is EUROSPACE. Composition includes 146 aerospace companies. In addition, there are eight United States companies listed as associate members. The organization was formed in 1961 upon the initiative of certain British and French industrial groups. The principal objective of EUROSPACE is the creation of a Western European industrial complex capable of providing governments, supranational bodies, and private interests with expert assistance and advice on space programs.

These three multilateral organizations have much wider interests than space communications per se. They are the international counterparts of the relation within the United States between NASA, the aerospace industries, and nonprofit corporations like RAND and are the West Europeans' organizational response to the Space Age. In contrast to the United States, where communications are owned by private companies and regulated by the Government, the principal communications agencies in foreign countries were government administrations.

The prime source of experience in international communications is the British General Post Office. This organization was one of the first to cooperate with the United States in space communications in the ECHO and TELSTAR experiments. As early as 1960, the Technical and Traffic Meeting of the Commonwealth Telecommunications Board discussed radio communications via earth satellites. Subsequently, the British Government invited its Commonwealth partners to a Commonwealth Conference on Satellite Communications held in London in April, 1962. According to the British Information Services:

The Conference recognized that it might be some years before satellite systems could become technically feasible, and, having regard to the research and development work being undertaken in the United States and elsewhere, that any commercial satellite system should serve as large a number of countries as possible and should have maximum flexibility. The Conference also acknowledged that satellite communications and submarine telephone cable systems would be complementary one to another, and had regard, throughout its discussion, to the projected submarine cable developments both by the Commonwealth and by other countries.[37]

While the British viewed satellite communications as a 1970's possibility, this did not mean that they were oblivious to the potential economic threat to cables. Shortly after the enactment of the Communications Satellite Act, Britain and Canada suggested that exploratory talks be held with the United States on the development of an operational system.[38] The Department of State concluded that talks would be beneficial, and from October 27 to 29, 1962, representatives of the British Post Office and Foreign Office, the Canadian Ministry of Transport and Department of External Affairs, and an interagency group headed by the Department of State met. The British and the Canadians "emphasized their desire to participate fully in the technical development, ownership and management of the system."[39] They thought, however, that another generation of cables could be laid before a space communications system would be operational. The United States took this opportunity to explain the purposes of the Communications Satellite Act, emphasizing the desirability of a single global system as opposed to competing national systems.

The British reported that they were expecting to submit a summary of these discussions to the Conference of European Postal and Telecommunications Administration (CEPT) to be held in Cologne in December. The State Department thought this move would only relate American ideas secondhand. Therefore, the State Department sent an American team to Europe to brief CEPT members prior to the December conference. During these meetings, and in the course of the conference, it became evident that the European countries were quite excited about participating in a joint system. The form of participation was not precisely envisaged at this time, however. More detailed consideration at the international level would have to await the formation of specific proposals by Comsat. The role of the State Department was to lay the groundwork for these discussions.

During these preliminary meetings, the principal Government representative, Gilbert Carter, perceived a pattern of internal competition within several European countries.[40] Past international telecommunications agreements had been negotiated on what might be called a nonpolitical basis. European telecommunications administrations had bargained with AT&T and other carriers about laying new cables or inaugurating radio services. Foreign offices were not involved in these negotiations. With the input into the traditional system of a means of communications whose economic and political consequences were far from predictable, it was natural for foreign offices to become involved. The disposition of questions affecting broad national interests could not be left for solution by traditional means. The telecommunications experts, on the other hand, wanted to preserve the old way of doing business. They believed that the diplomats would only create difficulties. This domestic alignment in certain European countries corresponds to the pattern within the United States where the carriers wished to preserve the old ways. To one trained in international relations, this pattern points to an interesting fact—the existence of a subculture in the international environment whose representatives have more in common with each other on a functional level than they do with nationals of their own countries who perform different tasks. But transnational cohesiveness of communications administrations did not serve as a barrier to participation by others. The introduction of satellite communications into the international environment could not be isolated, as was the introduction of telephone cables in 1956. Broader national interests were at stake.

In Great Britain, the Air Ministry, the Foreign Office, and aerospace companies were greatly interested in space communications. This broad range of interest was also the pattern in France, West Germany, and elsewhere. The rationale which had led to EUROSPACE and the talks leading to the formation of ELDO and ESRO worked here too. It was felt that in view of the technological leadership demonstrated by the United States, European interests would be best served if they could speak with one voice rather than many. The first concrete manifestation of a European regional approach came at the CEPT Conference in Cologne in December 1962. The Telecommunications Committee, one of the permanent organs of the Conference, set up a committee with the following terms of reference:

1. To study all the problems relating to the organized participation of all European countries desiring to do so in the establishment and operation of a single world network of telecommunication by satellite.
2. To establish particularly the basis of discussion to be held between the countries of the CEPT and the United States of America with a view to the possibility of establishing and operating a single world network of telecommunication via satellite.
3. To study the basis for a world organization for the management of such a network.[41]

The creation of this committee signified that Europe would take a regional approach to the United States proposal rather than negotiate a series of bilateral arrangements. In addition, it signified that the European countries accepted in broad outline the idea of establishing a single world system.

After the 1962 Cologne Conference, the Europeans proceeded at high gear to organize a European voice for space communications. An intergovernmental meeting in London in July 1963 entitled itself the European Conference on Satellite Communications (ECSC).[r] This Conference set up a committee structure consisting of a steering Committee, Organizational Committee, and a Space Technology Committee.[42] The committee which CEPT had set up in December became the telecommunications advisor to ECSC on technical matters. The other committees dealt with organizational, financial, and legal matters. Subsequently, in November 1963, the ECSC agreed that it "should be set up to provide to the extent possible, a counterpart to the U.S. Communications Satellite Corp."[43]

The existence of a European regional approach set the stage for definitive

---

[r]Also called CETS, derived from the French—Conference Europeenne des Telecommunications par Satellites.

discussions and negotiations to set up an operational global system. As a prelude to formal meetings a team of American officials from State, the FCC, and Comsat met with technical experts from the ECSC to discuss the outlook for satellite communications in the near future.[44] The first formal meeting was held in Rome in February 1964. No drafts were tabled at this meeting, but there was thorough discussion of the general principles which should provide the framework for the proposed system.

One basic characteristic of the system which was agreed to at the Rome meeting was the idea of having an interim system. It was felt that the establishment of a permanent international plan would be premature in light of the many unknowns—economic, political, and technical. One political variable was the role of the underdeveloped countries. Since the principal telecommunications users would be the main participants in the early period, it was felt that it would unduly complicate discussions to treat the newer nations as participants in the same category. In the words of Gilbert Carter:

The decision was to adopt a more limited approach and to try to work out among the principal telecommunications users the ground rules for a system, keeping in mind the necessity of providing adequate opportunities for all other nations to participate in or use the system in accordance with their desires and as it becomes technologically feasible for them to do so.[45]

Another idea that grew out of the Rome meeting was the agreement that Comsat should act as the manager of the system in behalf of all the other participants.[46] While ownership would be joint, Comsat would assure its primary place by being manager of the system. On the other hand, Comsat was not to have a free hand. The idea of an international steering committee to oversee developments was discussed.

As the Conference ended, there was considerable agreement on all but three major areas. These were (1) the duration of the interim agreements; (2) the allocation of ownership quotas, and (3) the voting procedure. The alignment of opinion on these issues put the United States against the ECSC. The Europeans wanted a short interim arrangement and the United States an agreement of considerable duration. The rationale of the American position was that with a longer interim period there would be more experience with which to negotiate a permanent arrangement. Also, the United States wanted a longer interim period in order to retain its initial leadership position. The Americans were afraid that the time needed to negotiate a permanent arrangement would be so great as to run counter to the mandate of the Communications Satellite Act.

The Europeans, on the other hand, hoping to increase their ability at the earliest possible time, pushed for a short agreement. The feeling was evident in certain European quarters that the communications satellite program might lead to American dominance in world communications of the future. For example, one member of the House of Commons asked the Postmaster General on February 26, 1964:

... there is now a growing feeling that, according to the trend of present talks, we shall finally end by starving the transatlantic cable of telegraphic communications from America and assisting Comsat to get off the ground, and that Britain will merely end up by renting a line from the Americans.

The Postmaster General replied:

... the Government's view is that the only way of preventing an American monopoly in this sphere is to join a partnership with the United States and other countries and so secure the right to influence the course of events. It is important to bear in mind that throughout the discussions there has been no indication that members of the Commonwealth have taken a different line on this.[47]

The idea that short interim arrangements would lead to an increase in the stature of Europe may have been wrong, however. As the system was to be global, the increasing number of non-European nations in the venture could align to work against the increase in European influence. Nonetheless, it was the position of the ECSC at the time of the Rome meeting that the arrangements should only last for about three years.

The ownership issue revolved around the amount of capital the Europeans wished to invest in the system. The greater the amount of a country's capital investment, the greater its percentage of ownership and thus, potentially, control and rate of return. Expressing their optimism on the prospects for satellite communications, the Europeans wished to invest more money than the United States thought they should.[48] The negotiations dealt with the figure of $200 million to capitalize the interim system. The Americans thought that this comparatively modest sum represented only a small part of the tremendous outlays for United States research and development. To base ownership only on the percentage of $200 million each country felt it could invest, would not clearly represent American leadership in the field. Therefore, the American team argued for substantial majority ownership by Comsat.

The voting issue concerned the formula by which the international steering committee (which came to be known as the Interim Communications Satellite Committee) would decide on various matters involving the management of the system. Since the United States was to have majority ownership of as yet undecided extent, it would naturally be able to control the majority of votes under the system of weighted voting that everyone envisaged as legitimate. The idea of one state, one vote was never entertained seriously because of the obvious functional differences in capability and responsibility. The problem lay in the amount of control that the Europeans would want. Important decisions would naturally involve weighted voting encompassing a greater number of votes than the United States alone possessed. Any other decision-making formula would destroy the concept of partnership. The issue, therefore, was not weighted voting but how much of a majority in addition to the votes of the United States would be required. The United States feared that the ECSC might vote as a bloc, and this would constitute a veto even under a weighted voting arrangement. The Europeans feared that, if they were not given some assurance on the criteria for making decisions on procurement, ground or earth stations, the space segment, etc., their standing as partners would be meaningless.

These three issues—the duration of the interim arrangements, ownership, and voting—dominated the discussions and negotiations following the Rome Conference in February. There were additional problems such as participation by the Department of Defense in the use of the international system and membership by the Soviet Union, but these did not directly impinge on the progress of events leading to the establishment of an interim system. Thus, these two problems will be discussed in later chapters.

The next meeting following the Rome Conference occurred in London from April 6 through 8. It was at this meeting that the American negotiating team advanced the idea of having two agreements—one on the governmental level and the other on the commercial level. The United States felt that the Europeans, Australia, Canada, and Japan would insist on an intergovernmental agreement rather than a purely commercial arrangement as has been characteristic in cable consortiums.[49] Therefore, it came prepared with a draft of a simple intergovernmental "umbrella" agreement.[50] This draft set forth principles to which governments would give a mutual guarantee. It was given tentative approval by the Europeans.

Following the April meeting a period of intense drafting activity began. In May, there were two meetings—one in London to compare drafts and the other in Montreal to study traffic statistics. The London meeting was not a negotiating session but an attempt to develop agreed-upon language for those points on which there had been no substantial differences. The Montreal meeting dealt with traffic projections. This was important because ownership in the interim system was to reflect the actual distribution of international telecommunications traffic. In order to measure the traffic, the 1963 International Telecommunication Union's projections for the year 1968 were used as a base. These statistics were reexamined in light of that part of the total international traffic which could be legitimately handled by satellite communications. It was agreed that the United States had more than 50 percent, but not how much more.[s] The meeting worked out a series of rough approximations to govern permissible investments by countries wishing to join the interim system.

The next meeting to discuss substantive differences occurred in London from June 15 through 16. At this time the issues of duration and ownership were resolved. It was decided that the planned Interim Communications Satellite

---

[s]Interview, Edwin J. Istvan, March 1, 1966. Part of the difficulty in estimating traffic was that communications from Hawaii to the continental U.S. are considered international because they are overseas communications.

Committee would submit a report no later than January 1, 1969, to all parties recommending what changes, if any, should be made in the Interim Arrangements. The report would be considered at an international conference, but, if there were no changes agreed to, the Interim Arrangements would remain in effect. The compromise gave the United States the assurance that there would be at least five years of experience before definitive arrangements would be suggested. The Europeans were satisfied because five years was not so long as to prejudice their position during the 1970s when they might have developed a greater technological base in satellite communications technology.[51]

The issue of ownership allocations was settled by giving Comsat 61 percent undivided interest in the system. It should be borne in mind that the system was the space segment of the total technology required. Ground stations, or earth stations, were not included as they were to remain in national hands. The total European participation, as well as that of the Canadians, Japanese, and Australians, was 39 percent.[52] A difficult facet of the allocation decision was how to arrange for the accession of new members to the Interim Arrangements. The Europeans were pushing to have the United States agree that, up to a certain point, all new signatories' shares should come out of the American quota. The United States successfully resisted this move, and it was decided that there should be a pro rata reduction of everyone's share as new participants decided to join.[53] The quota of Comsat, however, could not be reduced below 50.6 percent, thus assuring the Corporation a majority during the period of the Interim Arrangements.

At the end of June, only one important issue remained—voting. This was resolved at a conference in Washington which lasted from July 21 to July 24, 1964. It was agreed that on 14 important decisions, e.g. system choice, rates, earth station standards, placing contracts, the Interim Communications Satellite Committee would vote by a weighted majority of 12.5 votes above those controlled by Comsat. All other decisions would be by majority vote, but all parties committed themselves to try to arrive at agreement unanimously. In addition, it was decided that, if a decision on an important item concerning the budget, the placing of any contract, or the launching of a satellite were delayed more than sixty days, resolution could be made by a vote of 8.5 above those votes controlled by Comsat.[54] In order for this system of weighted voting to be acceptable to the United States, the ECSC had had to convince the United States that it would not vote as a bloc, otherwise the formula would give the Europeans a veto. While the smaller nations had wished to promote the ECSC as a regional group, the larger nations did not consider that their respective national interests would be best served by such a device.[t]

The voting formula is related to the crucial matter of contract awards. In Article X of the Agreement, it is specified that "the Committee and the

[t]Interview, Frederick John Dunn Taylor, Special Assistant to the Vice President, Technical, Comsat, British Post Office 1932-1965, March 21, 1966.

Corporation as manager shall . . . seek to ensure that contracts are so distributed that equipment is designed, developed and procured in the States whose Governments are Parties to this Agreement in approximate proportion to the respective quotas of their corresponding signatories to the Special Agreement." This particular provision was very difficult to negotiate. European industries had acted as lobbies through organizations like EUROSPACE to pressure their governments into negotiating for a plan which would assure them the chance to make substantial contributions to the procurement of the system.[55] The Europeans wanted an allocation for procurement on a national basis according to quotas of capital contribution.[56] However, the compromise provision allowed for geographical distribution only when competitive bids were comparable in terms of the best equipment for the best price.

It was obvious to most participants that substantial European contributions could come only in the 1970s. The initial system would be almost entirely American. Contracts had already been let with Hughes to send up a synchronous Early Bird satellite in March 1965.[u] In addition, Comsat had requested design contracts for an initial system from Hughes (synchronous), AT&T and RCA (medium altitude, random), and Space Technology Laboratories and ITT (medium altitude, controlled). A choice from these three alternatives was to be made in the fall of 1965—long before the Europeans would possess the level of technology required.[57] Nonetheless, the voting procedures and the procurement provision assured the Europeans that they could be partners in research and development as well as in the use of the system.

There remained only one unfinished piece of business, the negotiation on an arbitration agreement to be used in case of legal disputes. The drafting of this agreement was left to a group of legal experts, following the ratification of the principal agreements.[58]

Thirteen governments and the Vatican City initialed the "Agreement Establishing Interim Arrangements for a Global Commercial Communications Satellite System" on July 24, 1964. The Agreement was opened for signature on August 20 by the governments or their designated communications entities. Thus had a joint venture for space communications known as the International Telecommunications Satellite Consortium (INTELSAT) come into existence. Let us now evaluate INTELSAT according to the goals of United States foreign policy and in connection with the themes of the book.

### Conclusions

We have said that the Communications Satellite Act was a matrix for possible policies. As far as foreign policy is concerned, the events of the two years

---

[u]Early Bird was actually launched in April 1965. It is considered to be an experimental-operational satellite with approximately 240 2-way voice channels.

following passage of the Act gave some specificity to the general mandates of the Act. Let us elaborate what we previously called the foreign policy ambiguities of the Act,[v] to see what concreteness was given to them by the establishment of the Interim Arrangements. It will be recalled that there were five principal ambiguities: (1) the reasons for urgency in establishing the system, (2) the meaning of peace and understanding, (3) the nature of United States leadership, (4) the kind of participation by foreign countries, and (5) the relations between the Department of State and Comsat.

The necessity to establish an operational communications satellite system at the earliest practicable time was widely felt to be one of the principal reasons for passage of the Communications Satellite Act. The motives for urgency were both domestic and foreign. They were: (1) to promote American leadership in space in competition with the Soviet Union, (2) to encourage greater international peace and understanding, (3) to assure that an ongoing management team would bring to fruition the technology of space communications to fulfill the growing demand for international communications, and (4) to enhance the voting record of the 87th Congress. Unfortunately, the need for urgency also led the Government, on its own account, to circumvent the antitrust laws.

The speed with which an operational system could be established was conditioned by technical, economic, and political factors. Technically, it was predicted by NASA in 1962 that an operational system would not be feasible until at least 1967. Economically and politically, however, the system was well on the way to being launched in 1964. The EARC had allocated more than enough frequencies in 1963.[w] The establishment of the Interim Arrangements in 1964 met the political precondition of foreign cooperation. And, on the economic side, the Interim Arrangements and Comsat were more in danger of overcapitalization than lack of funds.[x]

In light of these speedy developments on the economic and political levels, could one reasonably conclude that the sense of urgency was false from the point of view of the Soviet threat? Speed in the solution of economic and political questions outpaced technological development. Hence, one might infer that the establishment of Comsat in 1962 could have been delayed a year or two. Suffice it to say that the creation of Comsat did not impede international negotiations and may have contributed to their early success. A similar agreement might not have been possible a year or two later.

---

[v]See Chapter 4.

[w]The final acts of the Conference entered into effect on January 1, 1965.

[x]On the domestic level, 10 million shares valued at $200 million were sold on a single day in June, 1964. The public received half and 163 communications companies half. The common carriers oversubscribed their share by $27 million which certainly evidenced their financial faith in Comsat. AT&T and ITT were the largest stockholders with $57,915,000 and $21,000,000 worth of shares respectively. Trading on the stock exchange started on September 8, 1964, and on September 17, Comsat held its first shareholders meeting and elected its first permanent Board of Directors. Thus, final ownership on the domestic level was not determined until after the international system was established.

One must conclude, then, that the dispatch with which international participation was accomplished reflected considerable merit on the capacities and the image of the United States in foreign affairs. Yet the opposition had contended that the ownership solution of the Act would detract from our image overseas. Thus, while the reasons for urgency in 1962 were ambiguous, the consequences of haste were not disadvantageous; they were surprisingly creative.

The establishment of the Interim Arrangements may have contributed to an increase in peace and understanding. The participants in the initialing of the Interim Arrangements all expressed great satisfaction and hope for the future.[59] But one would be foolhardy to confuse the euphoria of the moment with the substance of future international cooperation.

One might say that the Interim Arrangements were a victory for United States leadership in space. We have seen that there are several possible meanings of leadership,[y] but the Interim Arrangements may have maximized all of them. In the first place, the establishment of the Interim Arrangements was a policy triumph in relation to the Soviet Union. While the Russians sing the praises of international cooperation, their space program is nationally and bloc oriented. The Interim Arrangements proved that the United States space program is other-directed, not only experimentally as with the NASA programs, but operationally for commercial enterprise.

The United States helped create a system which was open-ended. The Interim Arrangements are flexible enough to allow developing as well as developed nations to join, albeit at a different level of ownership participation. In addition, the plan had provisions for its own modification should conditions change. Thus, the primary role of the United States was not rigidly written into the Agreement. American leadership in hardware was assured, but only until such a time as other nations could provide equipment on a comparable basis. Hence, United States leadership was promoted not in the harsh light of exploitive monopoly, but with a feeling for international understanding.

The reason the United States was able to negotiate an international system was directly related to its inclusive rather than exclusive approach towards participation. Participation is an ambiguous term which has several concrete meanings: (1) in research and development, (2) in manufacture, (3) in ownership, (4) in use and access to the satellites, (5) in management decisions, and (6) in program content. The United States negotiating position evolved from a series of propositions dedicated to the supremacy of Comsat to a proposal giving Comsat the primary role for an interim period. Other nations would have a say over procurement decisions, system design, etc. Comsat would be manager within the confines of the decisions of the Interim Committee. Access to the system would be assured for nonmembers of the Agreement, and on a nondiscriminatory and equitable basis. There would be no special rates for owners and nonowners as was at one time envisaged.

---

[y]See Chapter 4.

Only in control of message flow and program content would international control be lacking. This would be left to governments and individuals in keeping with past practice. This decision went against a conscious international program to change the flow and content of world communications to serve the goals of international peace and understanding. Instead, this goal would be achieved, if at all, by a laissez-faire approach. However, without active control, it is hard to see how the Interim Arrangements differ from the traditional cable consortiums in so far as the goals of peace and understanding are concerned. One is left with the conclusion that this goal of American foreign policy was more rhetorical than actual. Only by the unintended side effects of technological momentum could this goal be achieved.

One of the ambiguities of the Communications Satellite Act was the nature of the relationship between Comsat and the State Department. While the State Department had not been an active defendant of the original language of the President's bill, it was an active promoter of its own role after the legislation was passed. The Department's search for an international negotiating position initially provoked the lukewarm response of Comsat. But after a short period, Comsat-State relations improved noticeably. This undoubtedly contributed to the United States stance in the negotiations for the establishment of the Interim Arrangements.

The Congressional debate produced concern on the distinction between foreign policy and business policy. In practice, certain issues such as the type of management control involved both considerations. Congress reached a consensus on this issue by accepting its dual nature.

Let us now elaborate on the three interrelated themes of this book. The rather surprising conclusion that we must make about the rate of technological change is that it did not keep pace with political and economic developments. Frequencies were allocated, international arrangements begun, and capital contributions made two to three years ahead of the projected technological availability of a truly operational system. It had been argued in 1962 that an ownership solution was needed to assure a focus for establishing an operational system as soon as possible. The focus had arrived, but the technological lead time needed to produce an operational system lagged behind. It would be fallacious to argue that Comsat was needed to assure that technical work would continue, as this work was continuing in NASA and the Department of Defense. Rather, what we see is a case where political man was actually able to keep pace with technological man. Some observers might have political differences with the organization in the Government, Comsat, and the Interim Arrangements, but one could not deny that events on the political level had moved with great dispatch. A long lead time had enabled the United States to react to a technological innovation with time left over for thought and future planning.

The second theme concerns the patterns of interaction between foreign and domestic policy and is closely related to the third theme, the rationality of the policy-making process. It has been said that decisions are made legislatively when there are disagreements about goals with numerous alternatives open to participants relatively equal in power. These conditions of policy-making existed in foreign policy decisions involving frequency allocations and the establishment of the Interim Arrangements, although the United States was "more equal" than other powers. But the United States cannot be properly viewed as a cohesive participant during the 1963-1964 negotiations and deliberations. To say that the United States was the dominant power obscures the issue concerning the relations between Government departments, Comsat, and the carriers. Also, the proposition ignores the fact that the traditional pattern of behavior in international communications did not involve government-to-government relations in the sense in which diplomatic historians use the term.

What is interesting in examining the resolution of issues like voting and ownership quota is that the distance separating constituents of different nationalities was less than that separating some groups within the United States on closely related kinds of problems. While there was no insuperable obstacles in negotiating the Interim Arrangements, this was not the case for certain domestic political issues such as those involving earth station control, authorized users, and frequency allocations. Hence, we see a verification of Rosenau's line of reasoning.[z] The process of resolving some domestic issues can be more legislative and competitive than the resolution of certain issues on an international level.

In this connection, one is reminded of Gilbert Carter's testimony to the effect that there had been a subculture of international communications interests which had remained virtually isolated from the politics of the foreign offices, air ministries, and aerospace companies. Space communications technology punctured the isolation by introducing economic and political problems of high intensity. Not only were the consequences of management decisions both foreign and domestic from the American viewpoint, but they were also transnational. The negotiations for establishing the Interim Arrangements disturbed past practices and relationships in international telecommunications. The interests associated with this past pattern saw satellite communications as just another means of international communications, but the foreign offices and the aerospace companies disagreed. Hence, the past international subculture of communications carriers and government departments was modified.

Another example of this interpenetration of jurisdictions was the way in which the United States formulated a cohesive position from which to bargain with the European Conference on Satellite Communications. Consensus on the domestic level was lacking before foreign discussions took place. Consensus evolved from a process of accommodation involving the attitudes of foreign governments as well as the attitudes of the United States Government and Comsat. This points to the legislative character of the decision-making process. What one sees is another instance where the process of policy-making for space

[z]See pp. 6, 44.

communications technology blurred the traditional distinctions upon which policy had been based.

The fact is that technological change in satellite communications brought a new order of complexity into communications policy.[60] Technological change widened the area of choice and at the same time confused the old pattern of expectations. These consequences of technological innovation encouraged legislative decision-making and incrementalist rationality. Technological change discouraged long-range planning because the future became less certain and more open-ended.

Let us now turn to a consideration of another problem—space communications and national security policy. This subject also exhibits "muddling through" in response to technological innovation and other factors, in spite of the fact that the Department of Defense is often associated with a synoptic, cost-effectiveness approach to policy-making.

# 6

## Space Communications and National Security

The making of foreign policy cannot be divorced from the making of military policy. There are bound to be close relationships and crosscurrents, either conscious, subconscious, or hidden (to the public). Furthermore, the thrusts of both types of policy can be consistent or inconsistent, paradoxical or straight-forward. Military measures to increase defense posture can aid international peace through deterrence or lead to misunderstandings and possibly to war. In light of these inherent characteristics of military programs, it is not surprising that military and national security endeavors in space communications are often anomalous. Military requirements for space communications are both similar and dissimilar to civilian requirements. Military programs in space communications are both within the framework of American foreign policy as broadly expressed in the Communications Satellite Act and in opposition or competition with it. In this chapter, emphasis will be placed on clarifying the potentially paradoxical role of the national security programs for space communications and the relation of these programs to overall policy for space communications.[1]

Defense communications satellite programs have involved policy and organization problems in four areas: (1) within the Department of Defense where interservice rivalry and disjointed planning have often led to difficulties; (2) between the Department of Defense and other Government agencies such as NASA where coordination has been a growing requisite more or less successfully achieved; (3) between Defense and the business community where competition for contracts and Comsat's desire for traffic have had political overtones; and (4) between the Department and other governments where the paradoxical character of security programs created problems for the foreign policy goal of increasing international cooperation. Let us analyze these four issue areas in light of the following specific defense programs and proposals: Projects West Ford and ADVENT; the proposed Defense-Comsat system; the relations of the Department of Defense to INTELSAT; and the existing international military satellite systems.

### Projects West Ford and ADVENT

In Chapter 3, the discussion was left at the stage where the Department of Defense was working on two communications satellite projects—West Ford and ADVENT. Project West Ford was a passive communications satellite program conducted for the Air Force by the Lincoln Laboratory of the Massachusetts Institute of Technology. The project involved the dispensing into orbit of 400 million tiny copper dipoles 0.7 inch long and 0.0007 inch in diameter in a belt around the earth. As a reflector of communications beams, West Ford was impervious to military attack, but, as a possible hindrance to radio astronomy, the project created a great deal of opposition in scientific circles both here and abroad.

The Soviet Union tried to make considerable propaganda headway by picturing the United States space program as an outgrowth of military policy. To meet this opposition, the United States countered the criticisms openly rather than trying to preserve secrecy for national security reasons. Declassification of all essential matters in 1960 enabled the scientists of the world to examine the project objectively. The Space Science Board of the National Academy of Sciences reported that the program was entirely safe. Opposition was still heard on the international level, however, and, consequently, the President's Science Advisory Committee reviewed the whole issue. The possible hazards were again discounted in light of the facts, and a first launch of copper needles took place in October 1961. This was a failure and another try was made in May 1963, this time with success. According to the Space Science Board, the experiment was not harmful to either optical or radio astronomy.[2] All the dipoles in the West Ford experiment had decayed by 1968. Although West Ford was a very limited communications experiment, it symbolizes the type of conflict that can develop when the Department of Defense and the international scientific community collide.

The ADVENT project, in comparison to West Ford, was a broad-phased program designed to produce an operational capability. It was to involve active, synchronous satellites but because of program slippage, overoptimism, and increased cost, Secretary of Defense McNamara cancelled it in May, 1962. ADVENT spending had amounted to about $170 million of which only about $50 million was recoverable in the form of ground station hardware.[3] It was felt that a continuation of the program would have resulted in considerable further expense with questionable results. Therefore, the Defense Department switched programming from a synchronous satellite system to a medium-altitude communications system. Tight control over the development of this program was exercised to assure efficient allocation of funds in the service of a truly beneficial communication service. But, as the program definition phase reached its completion in August 1963, the Defense Department vacillated with the go-ahead signal for development. Instead, negotiations were undertaken with Comsat to see whether the Corporation might provide efficient service at acceptable rates.

## Defense Negotiations with Comsat: 1963-1964

The Defense Department discussions with Comsat are of interest to us because they occurred simultaneously with the attempt of the Corporation and the State

Department to establish international arrangements. Consequently, the idea that an international system of which the Defense Department as well as Comsat would be participants became an issue. The issue had ramifications on the domestic scene because it involved the questions of economy and Government use of private means of communications. On the foreign policy front, the idea of Defense Department participation in the Interim Arrangements had critical consequences for the type of image the United States wished to project in creating an international system. Hence, what was involved was a weighing of various economic, political, and technical factors to come to a decision which would affect the attitudes and actions of foreign governments, the ability of the United States to develop an effective, economical national security communications system, and the amount of revenues from Government communications to be distributed to the private sector.

Let us inquire why the Defense Department delayed making a decision which would have given a joint Philco/Space Technology Laboratory team a contract to build 60 satellites. In the fall of 1963, the Department realized that it had not objectively considered the alternative of a private system serving defense needs. This was heresy from the point of view of rational decision making, but was due to the following four factors according to Dr. Eugene G. Fubini, the Deputy Director of Defense Research and Engineering:

First of all, we had not properly separated the ground and the space segments. There were no questions in our mind that the Government needs to have control over its terminals . . . (but) the fact that the space segment and the ground segment do not go necessarily together had not been given sufficient weight.

The second question . . . was that we never felt and never thought that the commercial corporation would be interested in supplying the type of service we wanted and that would meet our requirements.

The third reason was a technical one. . . . It appeared to us almost incredible that it would be possible to come to a solution which would meet the requirements of the military for a few channels, with particular characteristics, while, at the same time, the requirements of a commercial corporation for a large number of channels with elaborate ground stations could be met. We were a little bit too hasty in our conclusions.

The final reason is that it is the policy of the U.S. Government that, if it is possible, we shall buy from the civilian sector what we need rather than do it ourselves, and if the reasons which have led us to the belief that we have to do it ourselves turned out to be incorrect, we should change our original idea.[4]

The Defense Department did change its original idea. In the hope that increased capacity at lower cost could accrue to the Government, McNamara requested Comsat on October 11, 1963, to indicate whether it had an interest in providing communications to meet military needs.[5] The Secretary of Defense made this request in the exercise of his role as agent for the National Communications System (NCS). The NCS had been created by President Kennedy on August 21, 1963.[6] By establishing a unified Government communications system, President Kennedy sought "to provide necessary communications for the Federal Government under all conditions ranging from a normal situation to national emergencies and international crises, including nuclear attack."[7] The Director of Telecommunications Management was given the responsibility of policy direction of the NCS, but, unfortunately, the post lay vacant and was to remain so until April 1964.[a] Hence, when the Secretary of Defense committed the NCS to negotiate with Comsat, it was not as a result of a coordinated interagency effort directed from the Executive Office of the President. In fact, the other members of the NCS were not notified. The Secretary did not even consult with the Defense Communications Agency, which had the responsibility of managing the NCS.[8]

Comsat was quite interested in the Defense Department's offer, and its reply led Dr. Fubini to say:

The corporation's research and development of a system suitable for the Government's needs can be expected to proceed at a pace which will permit the corporation to be in position to undertake first launch for such a system at least by early 1966 and to establish a worldwide operational capability by mid-1967.[9]

Let us consider what Defense requirements for an operational system were and how these differed from commercial requirements.

In listing needs for Project ADVENT, the Advanced Research Projects Agency (ARPA) had specified three kinds of communications needs: (1) international point to point, (2) ground to air and ship to shore, and (3) broadcast type to ground and mobile (airborne and waterborne) units.[10] According to the 1959 testimony of Roy Johnson, Director of ARPA, these three services were needed instantaneously as the existing cable and radio systems did not provide reliable communications under conditions of interference or jamming.[11] What was envisaged at that time was a military system of three synchronous satellites and four polar orbiting satellites to give complete global communications. The system would serve communications needs for crises. The bulk of defense traffic would continue to be sent through leased commercial facilities. This would be considerable, as the Defense Department then leased about 15 percent of the total capacity of the international common carriers.

By 1962, however, the Defense Department thought a medium-altitude communications system (MACS) was required, because the development of an effective synchronous system was a long way in the future. In 1959, the Department considered its needs to be immediate but technology would not oblige. In 1961, Brigadier General William M. Thames of the ADVENT Project stated that there was a gap between developments in weapons technology and developments in communications technology.[12]

The MACS system envisioned by the Philco/Space Technology Laboratories team would be a good interim system until technology allowed for a high-grade, permanent system. The MACS would provide fairly reliable communications to

---

[a]At this time Lt. Gen. James D. O'Connell of the U.S. Army, retired, was nominated to be DTM.

remote regions of the world. Hence, MACS was chosen as the system best able to meet national security requirements in the late 1960s.[13] The Department of Defense did not rely solely for future needs on satellite communications, however, but pressed AT&T to accelerate its laying of a new transpacific cable. Mr. James R. Rae of AT&T stated that "the economic comparison of ocean cables versus satellite systems in providing overseas communications is not yet clear. Costs of existing cables are known, but developments underway looking toward producing cable systems with transistorized repeaters capable of carrying more than 700 voice channels will greatly decrease the cost per circuit."[14]

Having chosen MACS as the system necessary for national security purposes, the Defense Department could only negotiate with the Corporation on the basis of its choosing this system. But in the fall of 1963, the Corporation had not yet decided which system it would require to meet commercial purposes. This choice would be made in the latter part of 1965. Thus, in all good faith, both parties had to arrange for a situation where their paths might cross. If Comsat, after comparing the synchronous and the medium-altitude alternatives, decided not to go ahead with MACS, then the Defense Department had two choices:

Choice No. 1 will be to tell the corporation "If you do not do it for yourself, do it for us, and we will pay you for it."

No. 2, tell the corporation we are taking over your R & D product at a reasonable price to be determined, and we will go ahead on our own.[15]

The realization that the needs of Comsat and the Defense Department might diverge pointed to the fact that their communications requirements were not symmetrical. The Aerospace Corporation, a "nonprofit" company which contributes scientific and technical support to the Air Force ballistic missile and space programs, undertook a study which pointed to the sources of the differences between national security and commercial requirements:

1. The require ments of a commercial communications system to earn high revenue as contrasted to the requirement of a military system to be everywhere available.
2. The requirement that a military system provide secure communications with as much freedom from jamming as possible and without requiring any cooperation except from military controlled ground stations.
3. The requirement that a military satellite system be as invulnerable as practical and that it degrade gracefully; that is, it is always better to have some communications system than none at all. The system must not fail catastrophically. These features, incidentally, are peculiar advantages of the medium-altitude random system, as distinct from the synchronous system.[16]

The Aerospace Corporation felt that, in negotiations with Comsat, the Defense Department was relaxing its requirements in order to come to an agreement.[17] Aerospace thought that a shared satellite system using one repeater with the same frequencies would undercut national security needs because of the problem of multiple access. By this was meant that commercial antennas would act as jammers for the military links. In addition, commercial frequencies were said to be unsuitable for military requirements and vice versa. Aerospace stated that if there were to be a shared system consisting of satellites with two repeaters, this would add so much additional weight to the satellite as to make launching highly expensive. In light of this analysis, it is not surprising that Aerospace suggested that Defense resume work on its own separate system.

Dr. Fubini was of the opinion, from the fall of 1963 through the spring of 1964, that a joint venture with Comsat would prove economical and mutually beneficial. But, whereas he first envisioned a shared system using satellites with one repeater, he now took the view that military needs could only be met under a joint system by satellites containing two repeaters.[18] Dr. Fubini did not underestimate the difficulties of establishing a joint system, but it was his considered opinion that such a system would save money. Although he could not estimate with precision the saving, he said it would be around $80 million, or 25 percent less than the costs to the Government of a separate system.[19]

The reason for the saving was based on the assumption that the Government would be charged rates based on an investment by Comsat of 40 percent of the total investment. Comsat had agreed to a 6-4 split in the investment base as indicating the extent to which Government research and development in satellite communications would entitle the Defense Department to a low rate base.[20]

Dr. Fubini disagreed with the contention of the Aerospace Corporation that the extra weight of a satellite with two repeaters would increase costs. Actually, with the new Titan III-C rocket, Defense hoped to save money as compared with the less powerful Atlas Agena. But, even with the use of the Atlas, Dr. Fubini contended that money would be saved through a joint system.[21]

The Defense Department also argued that none of its requirements for a military capability had been watered down. The only requirement which was altered because of the Comsat negotiations was time. Whereas Philco had been ready to go ahead with development in the fall of 1963, by carrying on the discussions with Comsat, Defense admitted that there would be about a year delay in the availability of the system.[22] In addition to this delay in time, which must constitute evidence for the view that the "urgent" needs of Defense were not instantaneously urgent, the Defense Department incurred costs of $400,000 per month to keep the Philco capabilities available.[23]

The case of the Defense Department, as represented by Dr. Fubini, was not only attacked by the Aerospace Corporation but by the House Military Operations Subcommittee and the Office of the Director of Telecommunications Management, as represented by Ralph L. Clark. Mr. Clark was of the opinion that the possible savings in a shared system were questionable, especially if there were to be two repeaters. He thought that the Defense Department had not yet come to grips with the technical problems. It appeared as if Defense wanted to agree first and figure later.[24] Government use of international and domestic communications comes to over one billion dollars per year, and the emphasis on savings through cooperation with Comsat, although commendable, seemed to Mr. Clark to miss crucial areas where substantial economies could be sought. These areas related to interservice coordination. Although this was the responsibility of the Defense Communications Agency, this agency did not have the operating capability to streamline military communications. The existence of three separate military communications systems was still the reality. Mr. Clark said:

The military departments are very large organizations and require a lot of communications to keep them operating. The worldwide military establishment is extremely large. It is going to take firm, consistent, and I believe long-range direction to alter the mix of systems and to evolve toward more efficient systems.[25]

The testimony of Mr. Clark raised the question of the motive of Government cooperation with Comsat. If economies of sharing were doubtful or minimal at best, why did not Defense seek economies elsewhere? And, conversely, what were other motives in seeking cooperation with Comsat?

Mr. Herbert Roback, the Staff Administrator of the Military Operations Subcommittee chaired by Representative Chet Holified (D., Cal.), suggested one rationale—a desire to subsidize the Corporation.[26] Dr. Fubini explained Defense's position by pointing to the Government's policy of not starting or carrying on any activity if it can be carried on as well by private enterprise. Mr. Roback replied:

Please let me interrupt. You are reading that tired old Bureau of the Budget circular, which really does not apply, when you know, in fact, that the Government has got a plant, an investment in communications plant that is something between $1 and $2 billion, and that the Government investment in this plant is bigger than the AT&T plant. Now, you are telling me that the Government is now adhering to a policy of not doing such things for itself. This is utter nonsense, and you know it.[27]

Mr. Roback was able to point to many soft spots in the Defense Department position. He indicated the fact that the proposed 40-60 split in the investment base was a matter for the FCC to decide, not just a matter for the Corporation and the Defense Department to negotiate. He asked whether the military requirements had not been "deferred or set aside in the interest of facilitating an agreement with the commercial corporation."[28] He seriously doubted whether Defense had the legal authority to negotiate the proposed agreement without Congressional authorization. To Mr. Roback, the whole endeavor lacked common sense, because the philosophy of the Communications Satellite Act was that the Government would construct its own separate system to fulfill unique needs.[29]

In May, after hearing twelve days of testimony, Mr. Roback summed up his case in a question to Dr. Fubini:

Restating now the case, you might say, against the shared agreements, tell us briefly wherein you disagree with the following points: No. 1, that the operating advantages are nil. This is from the standpoint of the Government. The advantages are nil. The economies are dubious. The growth potential is limited, and the time delay is substantial.[30]

Dr. Fubini replied that the operational disadvantages were nil and the economies and growth potential substantial. Only the delay in time was considerable, a matter of serious concern to the Department. But Dr. Fubini argued this way in May. By July 1964, Defense had scrapped its plans for a joint system with Comsat. Wherein lay the reason for the change?

The arguments against the proposed NCS-Comsat agreement from the point of view of dubious economies, lack of legal authority, and Government subsidy of private enterprise were apparently unconvincing to the Defense Department. According to the testimony, the Defense Department reinstituted its own program because it was impossible to assure military security requirements in a joint system composed not only of the Corporation but, in addition, foreign nations.[b]

Throughout the whole period involving the possibility of a Defense-Comsat agreement, the State Department had held an open attitude.[31] Although State, as a member of the NCS, felt that it should have been notified by Defense of the initial dealings with the Corporation, there was no indication of a recalcitrant attitude on its part. In March, State along with Defense and Comsat made a special trip to England, France, Germany, and Italy to acquaint these countries with the proposed agreement. But no itemized list of United States military requirements was presented to the foreign governments. The United States team felt it was inappropriate to "introduce into the international negotiations an item which was itself still the subject of negotiation."[32]

It seems that too much was left to chance, and that the Defense Department should have recognized the necessity for close collaboration with the negotiators of the Interim Arrangements, if it reasonably expected to work out a joint system. Possibly, the Defense Department was guided by the delusion that international arrangements could be negotiated without affecting the power of Comsat to control them.

The Director of Telecommunications Management, General O'Connell, called a meeting on July 8, 1964, to discuss whether the draft agreements for the international arrangements could be modified to take account of the objections

[b]Another interpretation could be that the Defense Department was convinced by the first arguments but was unwilling to lose face, and, hence, gave more weight than was needed to the international factors.

Interview, Daniel Fulmer, March 11, 1966.

of the Department of Defense. A consensus was reached that the Defense recommendations for changes were "of such a nature as to make them non-negotiable with the other signatories of the proposed international agreements."[33] As a follow-up to this meeting, Llewellyn E. Thompson, the Acting Deputy Under Secretary for Political Affairs, wrote a letter to General O'Connell in which he said:

...many countries would find it impossible politically to participate in a commercial system one component of which was reserved exclusively for the U.S. National Communications System. ...We believe that it would not be possible to negotiate these changes abroad, and that an attempt to do so might jeopardize our efforts to conclude agreements for a global commercial system.[34]

In the words of Mr. Roback, what followed was "a graceful degradation of the whole idea."[35] Understandably enough, Comsat was disappointed. It still looked to the possibility of working out an agreement with the NCS by means of a separate system if necessary.[36] At the Washington Conference to establish the Interim Arrangements, Mr. Welch suggested that the Interim Committee study the question of providing services to satisfy unique governmental needs of one of the participants.[37] The suggestion was approved by the Conference, but Mr. Welch did not specify a specific program or a concrete shared system.

## The Defense Satellite Communications System

By the summer of 1964, the Defense Department was put in the position of having to return to its initial contract with the Philco/Space Technology Laboratories team, which had been waiting in the wings at a cost of $400,000 per month and was eager to proceed with a concrete program. On August 8, the Secretary of Defense directed the Air Force to proceed with the interim system.[c] The plan was to orbit 24 satellites with the use of the Titan III-C boosters. The orbit would be semisynchronous as this would combine the advantages of coverage and being a moving target. The Secretary estimated that this system would last for three years, at which time it would be replaced by an advanced follow-on program. The advantage of this decision was that by delaying the system choice, the Defense Department said it was able to save $20 million. This saving principally reflected the substitution of the Titan III-C for the Atlas-Agena booster. Originally, Secretary of Defense McNamara had estimated separate system costs for the space segment at $60 million.[38] The consultations undertaken with Comsat had delayed the establishment of an interim system, but they had apparently saved $40 million.[39]

---

[c]Known as IDCSP, the Interim Defense Communications Satellite Project, as distinguished from ADCSP, the Advanced Defense Communications Satellite Project. The latter program was shelved as too costly and a more evolutionary program has been adopted.

Perhaps the goal of economy had been satisfied, but the procedures of cost effectiveness were not. However, once the Department of Defense returned to its closed arena of policy-making, divorced from the complications of merging its operations with government agencies or a private carrier, it could attempt to return to cost-effectiveness. The new system, initially known as the Interim Defense Communications Satellite Project (IDCSP) and changed in 1969 to the Defense Satellite Communications System-1 (DSCS-1), has worked quite well. Twenty satellites were operational at the beginning of 1971. They have performed better than expected and were not phased out by the follow-on DSCS-2 until mid-1971. The total development and production costs for the system through Fiscal Year 1970 were $352 million.[d] Whether there were any wastes or cost-overruns in the program is unknown to this observer. However, from the point of view of relating military requirements to civilian capabilities, it can be said that the military is insulated to the degree that it has spent about $23 million leasing satellite communications from commercial carriers compared to an average of $70 million per year using its own equipment in the Fiscal Years 1966 through 1970.[e]

### Foreign Relations Aspects of National Security Requirements for Communications: 1965-1972

Although the prospect of a shared system between the Interim Arrangements and the Defense Department was laid to rest in July 1964, the issues involved arose again in connection with (1) Defense Department use of INTELSAT for bulk traffic of an administrative sort; (2) the NASA-Comsat agreement concerning communications for the Apollo moon program; and (3) the authorized-user decision by the FCC in the summer of 1966. These matters have involved the following questions:

1. Government relations with private enterprise—the subsidy issue
2. The requirements and capabilities of the National Communications System
3. Foreign relations matters such as the promotion of United States leadership in space communication

In January 1965, a controversy arose as to whether the Defense Department would use its own system for bulk traffic as well as for urgent national security needs. Comsat was afraid that if Defense used its own system for all its traffic, the financial future of the Corporation would be dim. As the Defense Department share of all United States overseas traffic amounted to 30 percent, it

---

[d]Letter from John H. Sullivan, Public Affairs Officer, Defense Communications Agency, to the author. January 20, 1971.
[e]Ibid.

is understandable why Comsat was worried that the Government would not support it. Comsat had made an offer to Defense after the demise of the shared system plan. The new proposal would have cost the NCS $50 million, but Defense turned this down as being half-again as much as the Philco system.[40] The Department did assure Comsat, however, that it would not put all Defense traffic through its own system. Dr. Fubini estimated that "95 or 90 percent of our traffic will not go through the military system."[41] However, as it turned out two years later, the Department was planning to have one-third of its needs met internally and two-thirds by lease from commercial carriers.[42] As trends actually developed, by January 1971, the Defense Communications Satellite System was using 44.2 million, two-way channel miles—62 percent of them leased from private carriers and 38 percent government-owned. Basically, then, the private carriers have not received the share of business they thought they would. For communications satellite service, in particular, the Department spent $22,746,360 through the commercial carriers in Fiscal Year 1970, while on its own DSCS-1, it spent $111 million.[f] In addition, DSCS-2 will have 1,300 full duplex voice channels compared with between five and twelve channels for DSCS-1. Furthermore, Defense funding for space communications will rise from $62.9 million for FY 1972 to $192 million for FY 1973. This dramatic increase is due to the start of a new mobile communications satellite system called Fleetsat or the Fleet Satellite communications system.[43] This increased capacity and expenditure makes it likely that the share of business going to the private carriers will decrease even further in the future.

The interest of this distributive question lies in the evaluation of the importance of Governmental assistance to Comsat as a means of promoting the foreign policy goals of the Communications Satellite Act. There is a paradox here relating to the different roles of the Corporation. Leaving aside pros and cons of the financial controversy, one can make the following policy-oriented observations. If Comsat is a private enterprise, it would be contrary to free enterprise philosophy for the Government to assist it, if the Government could provide its own services more economically. On the other hand, if Comsat is viewed as the chosen instrument of American foreign policy in space communications, it would be contrary to the national interest to refuse assistance to the Corporation without which its financial and political image and stature abroad might be damaged. As Comsat is a little bit of both, one can readily understand why the subsidy issue is both a matter of domestic policy and foreign policy.

This same general analysis applies to the criticisms which were voiced against the 1965 contract which NASA, through the NCS, made with Comsat to provide communications for the Apollo program. The House Military Operations Subcommittee was again a forceful political interest in this issue.[44] The Senate Committee on Aeronautical and Space Sciences also dealt with the program.[45] Concerning the subsidy issue, Congressman Fernand J. St. Germain (D., R.I.) observed:

---

[f]Letter of John H. Sullivan to the author, January 20, 1971.

This is supposed to be a public corporation, and as I recall, the logic behind this and the arguments behind this were we want this to be a public corporation because the Government shouldn't be in private business, and here, lo and behold, the Government, one agency in itself, is going to buy 45 percent.[46]

Congressman William S. Moorhead (D., Pa.) asked Dr. Seamans of NASA whether, when the opportunity to get a regular customer from the Department of Defense had disappeared, Comsat had not "turned to NASA to bail them out of a hole?"[47] But Comsat had not requested the Apollo communications program (NASCOM). NASA, within the framework of the NCS, had examined all the alternatives and had on its own requested the manager of the NCS to contact Comsat.[48]

The Government had documents to prove that its choice was made after all due deliberation—as distinguished from the Comsat-Defense case. The House Subcommittee was especially concerned about the real economies involved, however.[49] Mr. Roback observed that the Communications Satellite Act provided that a separate Government system could be created in the national interest and that one of these interests was economy.[50] Mr. Roback was especially concerned that a large part of the decision-making process was classified. The feeling of the Subcommittee seemed to be that too much secrecy might cover up hidden purposes. The Subcommittee was willing to accept Government subsidy of Comsat, if this would strengthen its hands at the international bargaining sessions involving the setting up of definitive global arrangements for satellite communications after 1969. This would be a legitimate use of the taxpayers' dollars. Comsat could show its foreign partners that it did not depend fully on their cooperation to remain in business. This might strengthen Comsat's bargaining position by giving the Corporation an independent leg to stand on.[g]

The Military Operations Subcommittee was willing to go along with a subsidy if it were to serve this purpose. But the Subcommittee felt that the secrecy surrounding the NCS-Comsat negotiations should be removed. The use of taxpayers' money should be the subject of public debate.[h] But public debate might undermine the purpose of the contract, if it was in reality to subsidize Comsat. Foreign governments would recognize that an artificial strengthening of the Corporation was the reality. Perhaps this was the reason for the Government's secrecy. On the other hand, the Government provided evidence that the contract with Comsat was the most efficient and economical alternative available.[51] One can only conclude that the outside observer is left with a partial picture. However, the controversy points to the relationship to which we want to draw attention—the interrelation of the subsidy issue and foreign relations.

[g]This point was explained to me in interviews with Daniel Fulmer, February 23, 1966, and Edwin J. Istvan, March 1, 1966.

[h]Interview, Daniel Fulmer, February 23, 1966.

From the foreign relations angle, the NASCOM decision is interesting in that one sees international cooperation with a system which is at the same time serving U.S. Government needs. One reason the Defense Department-Comsat negotiations failed was because foreign nations were lukewarm to the idea of having a joint venture, but, in the case of NASCOM, a similar reluctance has not emerged. The explanation lies in the fact that use of a system serving the needs of astronauts has less political overtones than use of a system which serves confidential and urgent military needs. As it developed, the NASCOM system, which became operational in late 1966, has served both Apollo communications requirements and regular international traffic needs. The international system also handles military traffic of an administrative sort, but international cooperation involving urgent military needs has been restricted to United States arrangements with its allies. In September, 1966, an agreement was reached whereby Britain would acquire satellites in its own right as part of IDCSP. This program is known as SKYNET and began operations in November, 1969 with a successful satellite launch. However, a second satellite in the project failed in August, 1970. There is also a NATO communications satellite capability based on two satellites, one launched in March, 1970, and the other in February, 1971.

The third and most important problem to arise since the demise of the joint Comsat-Defense Department proposal concerns the authorized user issue. Section 305 (b) (4) of the Communications Satellite Act authorizes Comsat "to contract with authorized users, including the United States Government, for the services of the communications satellite system." One interpretation of this section would allow direct Comsat-Defense Department contracting with no necessity for FCC approval. This interpretation was that of the drafters of the legislation according to Edward Welsh.[i] Another interpretation would require the FCC to determine who were authorized users and under what circumstances the Government would fit in this category. The Defense Department proceeded on the basis of the first interpretation when, in January, 1966, it notified Comsat of its need for thirty voice circuits in the Pacific.[52] The FCC operated on the basis of the second assumption when it instituted in the spring of 1965 a public inquiry *In the Matter of Authorized Entities and Authorized Users Under the Communications Satellite Act of 1962* (Docket No. 16058).[53]

The Defense Communications Agency considered Comsat to be the best carrier and it notified the other international carriers of its traffic needs in the Pacific a full three months after it had notified Comsat.[54] It was not surprising that after proposals were received from five carriers at the end of May, Comsat's proved best. A contract was signed between DCA and Comsat on July 26. Perhaps it was also not surprising that the FCC, acting as the protector of the status quo, decided that Comsat could only deal with the Government directly "in unique and exceptional circumstances." On July 21, 1966, the Commission issued a "memorandum, Opinion and Statement of Policy" to this effect.[55] The

---

[i]Interview, Herbert Roback, September 6, 1968.

Commission believes that Comsat is "primarily a carriers's carrier" and "that if the Government or others were to obtain (service) directly from Comsat, there would be serious adverse affects upon the well-being of the commercial telecommunications industry and the general public it serves."

There was a clear difference of opinion, if not a conflict of interest, between the FCC and the Department of Defense. One way to resolve the difference would have been for the DCA to notify the FCC that the circumstances were "so unique and exceptional as to require service directly from Comsat." Although the DCA felt this way, this statement was not included in the contract. If it had been the FCC would have approved.[56] But the issue was pushed up to the Congressional level where the House Military Operations Subcommittee stepped in to recommend "that the DCA assign the Comsat contract to one or more American international carriers, based upon an across-the-board substantial reduction in charges for satellite and cable circuits in the Pacific area."[57] This solution was agreed to, and the saving to the Government through reduction in composite rates on all 128 cables circuits in the Pacific leased by the Department of Defense was greater than $6 million per year. This contrasts with the $1.6 million in savings which would have been realized by dealing directly with Comsat for 30 circuits.[58]

One can conclude that a new technology was slowed down to accommodate established interests, but, nonetheless, the results of incrementalism and the consensus approach produced economic benefits in the short run and avoided a sudden, destabilizing decision. The authorized user decision shows that the issue of cables versus satellites may never be explicitly resolved. There will be a mix of the two, although it will be slowly modified over time.[j] In fact, by 1972 the new office responsible for coordinating governmental communications, the Office of Telecommunications Policy (see Chapter 8), recognized that the heyday of decision-making through systems analysis had passed. Rational decisions in an area where one does not have all the facts and the facts change from day to day require day-to-day decision-making.[k]

## Conclusions

This chapter has provided information with which we can clarify (1) the rationality of policy making on complex issues, (2) the interaction between

---

[j]This conclusion is also supported by the FCC's TAT-5 decision of May 22, 1968. The Commission approved the laying of a new submarine cable to Southern Europe with a 720 voice circuit capacity. Cable and satellite facilities to this area are to be filled at a proportional rate. This institutionalizes a mix rather than letting market forces determine the best technology. In 1970, AT&T wanted to lay an additional cable, TAT-6, but the FCC and the White House called a halt to this move until a complete reassessment of policy could be made.

[k]Interview, Charles Joyce, Office of Telecommunications Policy, April 21, 1972.

the making of national security policy and foreign policy, and (3) the political context of technological change. First of all, we can see that the procedures which the Defense Department followed in making decisions on establishing a space communications system for national security purposes were not comprehensive. An initial decision based on an incomplete consideration of the alternatives was modified to take account of new factors. Furthermore, the decision to negotiate with Comsat could not itself be comprehensive, as future imponderables concerning the attitudes of foreign nations were not specified. Thus, instead of a decision being made by computers of systems analysts, one sees a decision being made over a year's time involving a process of mutual accommodation and clarification of variables, attitudes, and values. "Muddling through" produced a more rational decision in the sense of being technically, economically, and politically feasible.

Secondly, this chapter has indicated how the making of national security policy may interfere with the making of foreign policy. As national security policy by its very nature is aimed at potential enemies, foreign policy, at least in the case of communications satellites, was aimed at making friends. In the case of the Department of Defense's negotiations with Comsat for the use of part of the capacity of the international system, the paradox surfaced. Foreign nations would object to a system which placed priority on meeting the Defense Department's requirements. In turn, Defense objected to foreign control over any part of the system deemed necessary to fulfill defense requirements. In this case, Defense had to give way and it cancelled its negotiations with Comsat. It then established its own separate system. Subsequently, two international military programs were negotiated but these were with our military allies in NATO.

Defense had engaged in considerable planning to justify a joint Comsat-Defense system, but this planning was based on the assumptions of economics, not politics. Defense was more concerned with the division of labor than the division of responsibilities. But, with a broad perspective of the paradox of national security and foreign policy making and an attempt to coordinate Defense-Comsat discussions with Comsat-State-ECSC negotiations, the chances for success would have improved. Even then, the possibility of inconsistency may have killed the Defense-Comsat plan. The environments of policy making had basic dissimilarities, as well as certain parallel requirements.

In the third instance, this chapter points to the importance of the political context of technological change. Technological inventions are also social inventions occurring within the context of a certain set of group relationships and expectations. In the case of satellite communications to meet national security requirements, one sees a need not met by the pace of technological advance, as was originally expected. The ADVENT project, begun in 1960, was cancelled in 1962 after the Defense Department recognized that it was beyond the state of the art. Hence, urgent national defense needs were not met. DSCS-1 did not commence operations until 1966.

The expectation that technology would be developed faster than it was is an instance of overoptimism, perhaps stemming from a general opinion that all technological innovations are occurring at an ever increasing rate. In the case of the synchronous communications satellite development, one sees a technology which was not developed as quickly as other technologies. This is important information in that it points to the different life patterns of different technologies. It reminds us that we are not dealing with some abstract force called technological change.

Also, in connection with the subject of technological change, one sees that it did not alter the basic character of a perennial problem—the security dilemma in world politics. Instead, arrangements for organizing satellite communications were influenced by the basic political context of the Cold War. There could never be a single global communications satellite system. There would necessarily be at least two—INTELSAT and a Department of Defense System—and actually more.

# 7 INTELSAT and INTERSPUTNIK: American-Soviet Relations Concerning Space Communication

American foreign policy is often officially characterized as being guided by the desire to increase international peace and understanding. Unfortunately, these two goals are not always compatible. If understanding merely implies an increase in knowledge of other countries, then this could involve a growing awareness of differences in values and policies, thereby increasing the chances for hostility. Since the officials who enunciate these two broad goals are generally partisans of a relaxation of tensions, it can be assumed that by increased understanding is meant increased empathy and tolerance based upon the hope that increased knowledge will serve this end.

Another United States foreign policy goal is to increase American leadership. Concrete realization of this goal also involves ambiguity. Leadership in the sense of financial domination can be incompatible with increasing empathy or relaxing tensions. The goal of beating the Russians is compatible with one sense of leadership but incompatible with leadership in the sense of bringing about a detente between the two superpowers. The goal of a single global communications satellite system is not capable of realization if leadership in the space race is an overriding goal of American foreign policy. Another goal—that of increasing foreign participation in INTELSAT—is most likely incompatible with the goal of promoting the image of American leadership, for it is unlikely that the Russians would cooperate with INTELSAT as junior partners.[1] On the other hand, if promoting United States leadership means increasing American's stature in a competitive game whose outcome will have no effect on the national security of the Communist countries, it is then possible that the goals of leadership, a single global system, and increasing peace and understanding may be inclusive. But what this analysis shows is that the clarity of the goals of American foreign policy leaves something to be desired.

Let us ask ourselves to what extent periods of detente or confrontation in American-Soviet relations have been reflected in relations between the two superpowers in their exploitation of satellite communications. Relations between the United States and the Soviet Union in space communications have occurred within the larger context of the prolonged cold war, or lukewarm peace. They have not occurred in isolation but are an integral part of the overall pattern of relations between the two countries, and, in particular, they are a part of the pattern of interaction of the two nations' space programs. While it is not our intention to examine American-Soviet interactions in depth, viewing the patterns in space communications as part of the larger picture will provide greater understanding than an analysis which limits itself to describing space communications developments in particular.

In order to clarify the characteristics of the overall pattern of relations between the United States and the Soviet Union, it will be helpful to bear in mind the following distinctions. Relations can be cooperative, competitive, or conflicting.[a] These three concepts can be viewed as points on a continuum describing attitudes and behavior of nation states towards the values they would like to maintain or promote. The means of distinguishing cooperation, competition, and conflict is to see whether the ends pursued by a state are generalizable or exclusive. Generalizable goals can lead to cooperation between nations, while exclusive goals can lead to conflict. Unfortunately, the pursuit of common goals can also lead to conflict behavior; states may misinterpret each other's intentions.

Competition is a point on the continuum described by the existence of physical and/or verbal behavior in pursuit of goals which are neither purely generalizable nor exclusive. This situation exists when the prize at stake can be divided up without affecting the basic security of the states concerned. For instance, the United States won the race to land a man on the moon, but this was not a threat to the existence of the Soviet Union. This is why the moon program competition is often referred to as a race. It is more like a game than a fight to the finish.[2]

With these analytical distinctions in mind, let us now turn to a consideration of American-Soviet relations concerning space communications. The aim is to gauge the effect of the multifaceted American foreign policy for satellite communications on relations with the Soviet Union.

### Soviet Attitudes and Plans for Satellite Communications

The Space Age is also the Missile Age; it started that way in the most chilling years of the postwar period. The Soviet Sputnik signified not only the beginning of a competition in space but the existence, or so it was believed, of a missile gap accompanied by an increase in international tensions which have fitfully waxed and waned since the Cuban missile crisis.

The Space Age is a paradoxical age. There have been many proposals and policy statements enunciating the intention to exploit this new environment for the sake of international peace and good will. On the other hand, there have been a multitude of statements which indicate that the United States is in a race with the Soviet Union and that the outcome will affect American national

---

[a]I am indebted to the late Quincy Wright for this conceptualization explained during his International Politics Seminar at Columbia University, 1962-63. Professor Wright also added to this threefold conceptualization the pattern of coexistence—not really a relationship but a description of mutual noninterference. The following description of these concepts is my own.

security and the ability of the nation to survive. Practice, as distinguished from policy, has also been characterized by cooperation, conflict, and competition.

The Russians took an early interest in space communications. As early as 1957, a Soviet scientist predicted that the Soviet Union would develop the capability to broadcast television internationally via satellite.[3] In January 1959, the Technical Council of the Ministry of Communications proposed to include in the upcoming Seven Year Plan "the basic measures for realization of a cosmic television program retransmitter" to broadcast in color.[4] In August 1960, Robert Hotz reported that Soviet scientists were centering their attention on synchronous satellites.[5] In June 1961, the President of the Soviet Academy of Sciences stated that a high priority was being given to space communications. He wrote that "the use of communications, and satellites, and of satellites for relay services would revolutionize communications and television services."[6]

While the Soviets elaborated an idea, they did not officially set a time to attain this capability. While it is normal for the Soviet Union to broadcast the particulars of plans that will be accomplished "relatively soon," long-range goals are ambiguously, but confidently, projected.[7] Satellite communications were in this latter category in 1961 and 1962, while the Vostok flights of Majors Gagarin and Titov were not. Nonetheless, space communications are definitely part of the Soviet space program. However, the emphasis has differed from that of the United States. The Russians have talked more about television broadcast satellites than they have about point-to-point relays of telephone and telegraph traffic.[8]

This emphasis points to the fact that Soviet participation in traditional international communications amounts to about 1 percent of the total or less, while the American share is over 50 percent.[9] Since communications follow and lead to trade, and since Soviet trade with other countries is minimal, it is natural that the Soviets do not envisage the new technology as serving traditional purposes. Rather they see it as serving novel purposes such as direct-broadcast TV. However, a considerable technological lead time is needed to achieve this capability, and thus, the Russian plans for international space communications were long-term in the 1960s, although they may be short-term in the 1970s.

We know that the United States has had different plans and different capabilities. While many in the United States foresaw the new technology as revolutionary in the long run, in the short run the general expectation has been that space communications would meet the growing demands for overseas circuits to serve business and Government needs. In addition, there has also been a strong motivation, especially in 1961 and 1962, to establish American leadership in space and use communications satellite successes as a means to recoup lost prestige due to Sputnik. Thus, by August 1962, the United States had conducted five successful communications satellite experiments and had authorized the creation of Comsat. In 1964, INTELSAT was established. It was not until 1968 that the Soviet Union and seven other Communist states submitted a draft agreement to establish their own international communications satellite system, INTERSPUTNIK. However, in that same year, INTELSAT had increased its membership to 63 nations and, by the summer of 1969, INTELSAT operationally became a global system with satellites serving the Atlantic, Pacific, and Indian Ocean basins.

One may conclude that the United States has demonstrated its leadership in space communications through the 1960s and early 1970s, but Soviet plans may promise a more energetic exploitation of the new technology in the late 1970s. Before making reasoned conjecture on the future, however, let us estimate to what degree developments have been cooperative, competitive, and conflicting since the beginning of the Space Age.

## Cooperation

There have been halting steps towards cooperation in the space programs of the United States and the Soviet Union. These steps have involved cooperation in scientific research, meteorology, and manned moon ventures in addition to communications. On the legal level, they have involved the Outer Space Treaty of 1967,[10] the 1968 Agreement on the Rescue of Astronauts and the Return of Objects Launched into Outer Space,[11] the 1972 Convention on International Liability for Damage Caused by the Launching of Objects into Outer Space, and several noteworthy resolutions in the United Nations.[12]

The motif for cooperation has been set by the continued pronouncements by both countries that their space programs are peaceful. The International Geophysical Year (IGY), which lasted from July 1, 1957, to December 31, 1958, was the first example of the reality of international cooperation in space. But the IGY was itself a part of the seamless web of cooperation and competition. The Soviet Sputnik of October 4, 1957, was launched as part of the Russian program for the IGY, although its impact was felt beyond scientific circles. While attention since has been focused on the rivalry in space, the aftermath of the IGY has seen considerable cooperation in the setting up of the Committee on Space Research and the programming of many international experiments such as the World Magnetic Survey and the International Year of the Quiet Sun. However, these examples of cooperation are more in scientific *research* than *applications* like communications.

Cooperation in the applications of space technology for the benefit of mankind was the subject of an exchange of letters between President Kennedy and Chairman Khrushchev in February and March of 1962. The climate of opinion in which these letters were sent gave promise to more meaningful cooperation between the United States and the Soviet Union. In December 1961, the two nations had been able to resolve their differences and agree to forming the UN Committee on the Peaceful Uses of Outer Space. On February

20, 1962, Lt. Col. John Glenn, Jr. successfully orbited the earth putting the United States on a par with the Soviet Union. On February 21, Chairman Khrushchev sent President Kennedy a letter of congratulations which contained a statement pointing to the great benefits mankind could expect from space endeavors, if the two nations would pool their efforts. On March 7, the President responded and suggested collaboration in several specific areas—weather, tracking, mapping the earth's magnetic field, communications, and space medicine.[13] The President did not limit cooperation to these areas but indicated his willingness to discuss additional opportunities.

On March 20, the day after the Committee on the Peaceful Uses of Outer Space began hearings, Chairman Khrushchev replied by agreeing in principle to the areas of cooperation mentioned by the President. In addition, he proposed that agreements be concluded on the subject of assistance and rescue operations involving space ventures and on important legal problems of outer space.[14]

To follow through on a portion of these proposals, Dr. High L. Dryden of NASA and Dr. Anatoli A. Blagonravov of the Soviet Academy of Sciences met together to work out concrete programs. These meetings resulted in the recommendations of July 8, 1962, for cooperation in three areas—the establishment of a global weather satellite system for the benefit of all nations, the compilation of a map of the earth's magnetic field with the aid of satellites, and the improvement of communications by means of satellites. On December 5, 1962, the two governments announced in the United Nations that they had reached agreement on the recommendations,[15] and on August 16, 1963, NASA announced that an agreement had been reached on implementing the program.[16]

As far as communications were concerned, the agreement provided that experiments utilizing the ECHO II passive satellite were to be undertaken. In fact, over thirty such experiments have been conducted. According to James E. Webb, the former Administrator of NASA, "the project was carried out in a generally satisfactory way, although there were some areas in the Soviet Union's technical procedures which we feel could have been designed to give more useful results."[17]

This kind of cooperation has been characteristic of other space endeavors concerning both the United States and the Soviet Union. The language of joint interests confronts the reality of sovereign states with separate interests. What cooperation has occurred has not been in the form of joint ventures such as was suggested by Presidents Kennedy and Johnson in connection with the moon program.[18] Rather there has been cooperation in the form of coordination. This coordination has occurred in areas where the results of mutual understanding will have an inconsequential effect, if any, on the military or prestige balance.

These limited successes in the coordination of scientific research in the 1960s have continued into the 1970s with the October, 1970 agreement concerning compatible docking systems for manned spacecraft—a necessity if the Agreement on Assistance to and Recovery of Astronauts was to have any concrete meaning—and the January, 1971 agreement concerning exchange of lunar samples, improvements in the exchange of weather satellite data, coordination of meteorological rocket soundings, exchanges of data on the natural environment, the joint scientific investigation of near-earth space, and the exchange of medical information on man's reaction to working in space.[19] Notably lacking from this last agreement was any proposal for cooperation in the field of communications. According to George M. Low of NASA, this was because "we were unable to achieve an understanding of the need for structuring an experiment to achieve mutual benefits."[20] As there had been a cooperative experiment in the early 1960s using the passive satellite, ECHO II, one may raise the question as to why such joint experimentation has not continued. It seems reasonable to suggest that cooperative research which appears to be successful in developing knowledge of possible applications in the future will, by the very nature of its success, tend to undermine cooperation in this area in the future. This conclusion follows from the proposition that agreements involving research are likely to be fruitful while operational agreements involving areas of competition and conflict are not likely to be negotiated. Only if operational programs involve humanitarian concerns, as is the case with the Rescue of Astronauts, will there be a likelihood of establishing joint endeavors.

While cooperation between the two super powers has been limited, it has not been inconsequential. Thus, it is wrong to conceptualize the space relations between the United States and the Soviet Union as purely those of competition and conflict. The Dryden-Blagonravov Agreement, the EARC, and UN Resolution of 1962 were all products of a time span in which the Nuclear Test Ban Treaty came into force, the Soviets ceased jamming Voice of America broadcasts, and the United States and the Soviet Union proposed a resolution to the General Assembly to ban the orbiting of weapons of mass destruction. Thus, developments in late 1963 represented a detente in international hostility, although the underlying causes of conflict may still have remained.

But from 1964 through 1966 cooperative programs stagnated. In 1967 and 1968, there was a brief respite from the Cold War indicated by the Outer Space Treaty and the Agreement on the Rescue of Astronauts. Then from late 1968 to late 1970, when cooperation in space research was resumed, the American-Soviet competition in space increased. Parallel to this growing rivalry, evidenced by the INTERSPUTNIK proposal and the United States' success at landing men on the moon, relations between the superpowers in other areas reached new lows, as witnessed by the escalation of the arms race, the extension of the Vietnam War to Cambodia, the Six Day War, and the invasion of Czechoslovakia. However, paradoxically, the years 1970-1972 may be viewed as years of increasing cooperation and competition. The rivalry is illustrated by the above examples and such other factors as the increasing Soviet naval program, the American mining of Haiphong and increased bombing throughout Indochina, and the consequences of the India-Pakistan war. During the same period, summits were held in China and Russia (both covered by television through the INTELSAT system), a treaty was signed barring weapons of mass destruction from the ocean floor, and a draft treaty was agreed on to prohibit the production of possession of biological or toxic weapons. Let us now view the competitive side of this confusing and complex picture.

## Competition

During the 1962 meetings of the Legal Subcommittee of the Committee on the Peaceful Uses of Outer Space, the Russians objected to the United States Project West Ford and the operations of Comsat.[b] These objections seemed to render unlikely the conclusion of an agreement on legal principles governing man's activities in space. The Soviet draft of legal principles relating to satellite communications stated:

1. Cooperation and mutual assistance in the conquest of outer space shall be a duty incumbent upon all States; the implementation of any measure that might in any way hinder the exploration and use of outer space for peaceful purpose by other countries shall be permitted only after prior discussion of and agreement upon such measures between the countries concerned.
2. All activities of any kind pertaining to the exploration and use of outer space shall be carried out solely and exclusively by States; the sovereign rights of States to the objects they shall launch into space shall be retained by them.[21]

The first of these principles was directed against Project West Ford and was unacceptable to the United States, not because it wished to carry on this project or others without taking a proper regard for the opinions of mankind, but because it did not wish to subject national programs to a Communist veto.

The United States objected to the second principle because it undermined Comsat. By stressing the exclusiveness of state responsibility for and control of space endeavors, the Russians meant to downgrade the role of private enterprise or, as the Soviets would call it, exploitative monopoly. It is clear that the Russian principles witnessed the intrusion of the Cold War into what had seemed a promising move toward cooperation in the form of the Kennedy-Khrushchev letters.

Nevertheless, the end of 1963 was characterized by a spirit of agreement, as discussed above. The Soviet Union agreed to omit consideration of the principles they had so strongly urged in the Legal Subcommittee. The General Assembly was able to unanimously pass Resolution 1962 on December 13. But 1964 ushered in another period of Cold War competition.

The Soviet Union evolved a position extremely critical of Comsat and

[b]They also objected to Project Samos, a reconnaissance satellite program.

INTELSAT. They were slow in voicing this antagonism, however, and thus, most likely came to it after a period of long thought. In February 1963, the American Embassy in Moscow delivered a note to the Russians suggesting that it might be useful to have discussions on a commercial communications satellite system.[22] The Russians, however, considered a meeting premature. They also took this attitude during the EARC later in the year. But, in 1964, the United States had initiated negotiations on a concrete basis with various Western European and other nations. In February 1964, perhaps for reasons of intelligence, the Soviets expressed interest in consultations with the United States. They suggested that a meeting be held in Geneva, during the May and June meeting of the Technical Subcommittee on the Peaceful Uses of Outer Space.

The United States found this agreeable, as it is a mandate of the Communications Satellite Act to seek agreement with other nations. Hence, in June, an American team of Government and Comsat officials flew from the London meetings, concerning the drafting of the Interim Arrangements, to Geneva. This group briefed the Russian delegation headed by Dr. Blagonravov about the negotiations for establishing a global commercial system. The Russians asked a few questions concerning whether the EARC allocation decisions were definitive enough to warrant the engineering and construction of an operational system. This line of inquiry still reflected the Soviet view that operational arrangements were premature. Consequently, at the end of the meeting, Dr. Blagonravov read a statement which was essentially this:

This is all very interesting. We still consider that technically you are in an experimental phase. This is an American or U.S. inspired experimental program which you are embarking on. We are not really very interested in it at this time. We will continue to do some experimental work on our own. And perhaps at some time in the future we will get together and talk about the whole project again.[23]

This rather opaque and lukewarm statement did not reflect the ideological opposition to private enterprise which was to characterize Soviet attitudes later in the year. Perhaps the explanation for this lies in the fact that Dr. Blagonravov is a scientist who had concluded the agreement with Dr. Dryden concerning American-Soviet cooperation in space. By leaving the door open, Dr. Blagonravov may have expressed an attitude contrary to that of his political superiors. It would seem that the Russians might have much to gain from cooperation with the Interim Arrangements which would pool resources and cut costs. But for political reasons the arguments of economy often go unheeded.

In the fall of 1964, at meetings of the International Institute of Space Law and of the Committee on the Peaceful Uses of Outer Space, the Soviet Union and other Communist countries expressed strong opposition to INTELSAT as a device for transferring space communications into the hands of United States private capital. In Warsaw, Dr. I.I. Cheprov read a paper in which he contended that it was "the duty of lawyers imposed upon them . . . by resolutions of the UN General Assembly to see to it that communications by means of satellites. become 'available to the nations of the world as soon as practicable on a global and non-discriminatory basis.' (Res. 1721/XVI)"[24] Dr. Cheprov contended that the Interim Arrangements were a means of perpetuating a monopoly—Comsat. Not only would partners in the agreements be the victims of monopoly domination, but so would Americans themselves. "It is well known that private enterprise in the U.S.A. has a long experience of discriminating against Americans themselves, so how can we expect this enterprise to be unbiased and just on the international scene?" Dr. Cheprov quoted the opponents of the Communications Satellite Act as evidence against existing American policy.

These charges were repeated at the meetings of the UN Committee on the Peaceful Uses of Outer Space held between October 26 and November 6, 1964. The delegates of Communist block countries criticized INTELSAT as a profit system for the benefit of the undeveloped countries at the expense of the developing nations.[25] Secondly, they criticized the Interim Arrangements as a violation of the principle of sovereign equality due to the procedures for weighted voting. And thirdly, they criticized them as an attempt to create an organization outside the proper organizational and political framework of the United Nations and the International Telecommunication Union. These charges were couched in the traditional semantics of the Cold War. They were placed in perspective by the dispassionate analyses of the representatives of the United States and the United Kingdom.

Mr. Thatcher of the United States pointed out that a system based on profit was not necessarily exploitative. "Perhaps what we refer to as 'profit motive' in this country is not as far removed as it might first appear to be from what sometimes is translated from the Soviet press as being 'profit incentive.' " Mr. Thatcher went on to suggest that the Communist delegates might be mistakenly confusing participation in the INTELSAT with access. He said, "It is clear from the agreements themselves . . . that participation in these arrangements is open to all states members of the ITU and that whether or not a State is a participating member, access to the system is on a completely free and non-discriminatory basis to all states."

The Soviet charges against the Washington agreements relating to denial of sovereign equality through weighted voting and the creation of an organization divorced from the UN framework were not dealt with by Western delegates at this time. However, several considerations in this regard may be mentioned. First, the principle of sovereign equality in some arrangements is not incompatible with the principle of weighted voting in others. Political treaties are usually based on sovereign equality, although even here the Security Council can be viewed as an exception. Functional arrangements may involve special formulas reflecting unique situations and responsibilities, as with the ITU, INTELSAT, or the Common Market. Thus, the presence of weighted voting does not per se lead to exploitation. It may reflect considered division of responsibilities.

Secondly, INTELSAT is not divorced from the framework of the UN or the ITU. The Extraordinary Administrative Radio Conference of the ITU set the technical framework for the formation of a system of communications satellites. And the UN does not discourage, but, in Chapter VIII of the Charter, encourages the formation of regional arrangements whose principles harmonize with those of the Charter. It is somewhat strange to hear the Soviet Union criticize the United States for carrying on operations beyond the fringe, when the Russians themselves often appear as the consummate devotees of self-insulation. But the tactic of calling for increased international cooperation for others, while reneging on international collaboration itself, is a familiar Soviet practice. In fact, what one sees here is another form of competition—competition to see who cooperates the most.

Since both the United States and the Soviet Union continually pronounce their cooperative intentions, it can be interpreted as a mark of prestige to have a better record of international collaboration. For instance, James Webb has said:

Despite its protestations of peaceful interests, the Soviet space program can show no comparable (to our own) engagement in cooperative relationships.[26]

To overcome this weakness, the Russians have proposed their own international satellite communications system, INTERSPUTNIK. On August 5, 1968, the USSR, Bulgaria, Cuba, Czechoslovakia, Hungary, Mongolia, Poland, and Romania submitted a draft agreement for INTERSPUTNIK to the United Nations.[27] This proposal took advantage of several apparent weaknesses in the INTELSAT approach. In the first place, the only members of INTERSPUTNIK are states and the voting procedure is one state, one vote. This approach would obviously appeal to many countries, for it contrasts favorable with the image of INTELSAT as being dominated by a private, United States corporation. Secondly, the Preamble to the draft contemplates the establishment of a system which would provide direct broadcasting from satellites. Such a system could bypass the need to establish expensive earth stations to relay signals from satellites in space. Thus, if this is what is implied, INTERSPUTNIK could greatly aid the less developed countries by enabling them to circumvent the need to establish a costly infrastructure for communications. Thirdly, the INTERSPUTNIK draft agreement allows for more than one international communications satellite system, whereas one of the main criticisms of INTELSAT is that the signers of the Interim Arrangements committed themselves to a single international system.

Unfortunately for the Soviet Union, there were and are several drawbacks to the INTERSPUTNIK draft. The first was situational. The same month the eight Communist states proposed their own system, the Soviets launched their invasion of Czechoslovakia. The possible propaganda benefits of INTER-

SPUTNIK were consequently overshadowed by the callousness of their behavior towards their own ally.

Another sort of drawback to INTERSPUTNIK involves an analysis of the ambiguities in the draft proposal. There are three principal ambivalencies. One concerns who will own the system. Article 3(2) states that the satellites may be either the property of INTERSPUTNIK or leased to INTERSPUTNIK by its member states. As Stephen E. Doyle writes, this provision "could be construed as a Soviet effort to ensure that the USSR would be able to provide at least some of the space segment requirements of the international system to the organization on a lease basis with the USSR collecting the rent."[28] It follows, that, in contrast to INTELSAT, INTERSPUTNIK could be a coordinating, umbrella organization rather than one which is the sole owner of the space segment of the international system. On the other hand, parallel to INTELSAT, Article 3(4) of the draft states that "the ground complex shall be the property of the States which have constructed it in their territory."

A second ambiguity in the INTERSPUTNIK proposal concerns the lack of any requirement that there will be nondiscriminatory access for all users, a requirement which is clearly stated in the Preamble to the INTELSAT agreement. On the other hand, Article 10 of the Communist draft states that "the distribution of communications channels among states members of the organization shall be made on the basis of their need for communications channels." However, in paragraph 2 of this same article, it is stipulated that a state must pay for its channels at fixed rates, surely a strange, albeit understandable requirement for a Communist organization ideologically committed to the proposition "From each according to his ability, to each according to his needs."

The third principal source of ambiguity in the INTERSPUTNIK draft concerns the organization's executive and governing structure. As voting occurs on a one state, one vote basis, and as a two-thirds majority is required for action, it is not clear how action will come about in the event a minority wishes to stall an initiative backed by a simple majority.[29] In addition, there is no international executive body as there is in INTELSAT, but there is a Secretary-General whose responsibilities seem to include a great deal of executive as well as administrative authority.

The third major drawback to the Soviet draft proposal for INTERSPUTNIK is that no states have taken the Soviets up on their offer four years after its submission. It is true that the draft treaty of 1968 has been superceded by a treaty in final form which was deposited at the United Nations on November 15, 1971, but this treaty had not come into force by mid-1972 and, furthermore, INTERSPUTNIK has not begun actual operations. The Soviet Union has established a domestic system, ORBITA, which is quite successful and extensive.[c] But their interest in international space communications has not been pressed with as much vigor as their programs in space exploration and the

---

[c]This system was initiated in 1965 and is based on Molniya (Lightning) satellites, which have a capacity of about 2,000 voice channels.

exploitation of near-earth space.[30] However, the Russians have plans to launch a synchronous communications satellite, to be called Statsionar I, over the Indian Ocean. This position will allow it to cover populated areas between England and Japan. Also, the Soviets have helped construct earth stations abroad (in Mongolia, Cuba, the U.A.R., and Mali).[31] It is therefore possible that INTER-SPUTNIK may yet get off the ground; but it is unlikely that the system will be able to compete on a point-to-point basis with INTELSAT. The only potential threat to American leadership through INTELSAT could come in the late 1970s through the establishment of a direct-broadcasting system.

In summarizing the competitive features of the American-Soviet relationships regarding space communications, one must emphasize that the United States has a commanding lead in exploiting the existing technology for purposes of international communications.[d] However, the character of American leadership is such that it damages the prestige but not the security of the Soviet Union. The Russians are not impervious to the prestige competition, but, through 1971, they have limited their responses to ideological thrusts and limited international experiments.[32] However, one must remember that we are interpreting the competition at only one stage. Perhaps the Russian plans for INTERSPUTNIK will present a real challenge to American prestige in the late 1970s. The INTELSAT program is principally aimed at fulfilling the rising demand for communications between industrialized countries. If the Russians were able to establish a direct-broadcast TV service to the newer nations, they might have at their disposal a powerful instrument of progaganda and mass communications.[33] Whatever the likelihood of a Soviet move in this field may be, one can say that American policy makers are not blind to the possibilities. In fact, there has been considerable discussion of such a system in the United Nations and the Congress.[34]

## Conflict

Let us now turn to a consideration of the conflicting aspects of the American-Soviet confrontation in space communications. Here, attention should mainly be focused on the military balance. The Defense Satellite Communications System is an integral part of the United States need to link together its far-flung Army, Navy, and Air Force facilities.[e] Concurrently, one can assume that the Soviets are using Molniya satellites for military as well as civilian communications.

Viewed as part of the military establishments of the two nations, space communications systems have a potential for becoming part of a conflict between the two super powers. Most likely, in the event of war, efforts would be made to destroy or jam the communications networks of the enemy. On the other hand, in severe crisis situations, advanced satellite communications systems

[d]See Appendix A for charts listing U.S. and USSR launches.
[e]See Chapter 6.

can serve as part of the "hotline" between the two superpowers and thus allow for the possibility of more rational communications between them. Under an agreement signed by the United States and USSR in September, 1971, the two states "will notify each other immediately in the event of detection by their missile warning systems of objects that they are unable to identify or in the event of signs of interference with such missile warning systems or related facilities."[35] And in peace, or what is now a fairly stable deterrent situation, space communications systems can be considered as a means of strengthening the balance. By increasing the ability of both countries' early warning systems, space communications networks could increase the stability of the system of mutual deterrence.

The entire space program of each country can be viewed as part of a conflict for control of the new frontier, as were past explorations and exploitations of the sea and air; but, such a comparison seems far-fetched. To other observers, however, the space program seems to be a means for preserving America's very existence in the world. At times, former President Johnson spoke of United States space goals in this way. In a 1964 NASA briefing on Project Ranger, the President asked Dr. Pickering: "This is really *a battle for leadership and real existence* in the world, isn't it?" (Italics mine.)[36] Dr. Pickering agreed, and the following dialogue took place:

The President: In effect, the British dominated the seas for centuries and led the world, didn't they?
Pickering: Yes, Sir.
The President: We have dominated the air with leadership, and I think unquestionably have been *the leaders of the free world* since we established that *dominance*, haven't we?
Pickering: Yes, Sir.
The President: And the person that leads in space is going to have an equivalent position, isn't that true? (Italics mine.)[37]

This rhetorical line of questioning established President Johnson as a somewhat ambiguous interpreter of the national space program. The program is related to "a battle for leadership and real existence," leading the free world, and "dominance." Furthermore, the former President contended that leadership via the manned moon program "is essential for our civilization."[38] Such an assessment of the Apollo program and the space program as a whole connotes a mixture of competition and conflict weighted towards conflict. The emphasis is towards a zero-sum game where one side wins and the other loses. Not only is leadership involved, but real existence. It is a question of dominance, not compromise.

The Russians have often taken this same attitude in relation to space. For example, in September 1962, it was reported by the Soviet Defense Ministry that in order to counter imperialist military ventures in space, the Soviets would have to retaliate.[39] In the same month, a Soviet book on military strategy was published containing a similar statement:

... The Sovient Union cannot disregard the fact that United States imperialists have subordinated space exploration to military aims and that they intend to use space to accomplish their aggressive projects—a surprise nuclear attack on the Soviet Union and other Socialist countries ...[40]

The tenor of these remarks indicates an implicit threat to enter into an arms race in space, if the provocation is great enough. But both super powers have avoided such a course. Military activity in each country appears, in light of the public documents, to be limited to communications, reconnaissance, and early warning systems.[f]

What one sees is a pattern of watchful coexistence which would only break down in the event of hostilities. This pattern applies to the entire space program of each country. There does not seem to be a valid point in talking of America's "real existence" being threatened by Russian activities in space—in communications, lunar projects, or elsewhere. The confrontation in space is more a race than a fight. One is reminded in this connection of President Eisenhower's talk at the Naval War College in 1961, where, speaking four months after President Kennedy committed the nation to the landing of a man on the moon within the decade, he said, "I think to make the so-called race to the moon a major element in our struggle to show we are superior to the Russians, is getting our eyes off the right target, I really believe that we don't have that many enemies on the moon."[41]

### Conclusions

Let us conclude this chapter by estimating the varying degrees of conflict, competition, and cooperation which have characterized American-Soviet relations concerning space communications. In this connection, it will be appropriate to discuss briefly the clarity of American policy for satellite communications as regards its projection to the Soviet Union. In addition, it will be relevant to discuss how disruptive communications satellite technology has been in its effect on the strategic balance.

It is evident that satellite communications, as a subject of concern to the United States and the Soviet Union, has mainly involved competitive forays rather than cooperative ventures or outright conflict. In turn, there has been more cooperation in the sense of coordination than conflict over vital interests. The dominant theme has been competition marked by American success with INTELSAT and Soviet failure with INTERSPUTNIK, although the Russians have had success with their domestic system, ORBITA.

[f]One exception to this may be the Fractional Orbital Bombardment System (FOBS), which is not strictly speaking a space weapon because it only has a fractional orbit.

While the technical success of the United States projects in satellite communications is apparent, one might not be able to say the same thing about United States policy. This is because of the ambiguity of the policy. Important decision makers continually make statements reflecting a desire to dominate in space while, at other times, they promote an image of cooperation for the benefit of mankind. Such statements must be confusing to the Soviet Union. But, likewise, such statements are made by Soviet officials, thus increasing the chances for misunderstanding in the United States.

It is this observer's impression that the introduction of communications satellites technology has not altered or effected this syndrome of ambivalent interaction between the United States and the Soviet Union. Technological innovations in weapons systems have proved more profound in altering the relationship between the two super powers than technological changes in communications. The latter technology has thus far been absorbed into the dynamics of the Cold War system of mutual deterrence.

While satellite communications is not a potent instrument of foreign policy for either the United States or the Soviet Union, it is certainly a potentially significant tool. The success of INTELSAT[g] has demonstrated in a clear way the superiority of the American cooperative approach. But INTELSAT handles communications on a commercial level. Most interaction between peoples is commercial rather than diplomatic or cultural. However, the saliency of satellite communications as a tool of policy in helping the developing nations to develop their internal and external educational and entertainment, as well as commercial, communications networks should not be overlooked. This is what the Soviets have stressed, and the United States may be facing the challenge of this opportunity in the late 1970s.

---

[g]See Chapter 8.

# 8

## The Transition Between the Interim and Definitive Arrangements for INTELSAT

In this chapter, attention will be focused on the clarity and cohesiveness of American space communications policy since the establishment of INTELSAT in 1964. To structure this inquiry, we will first analyze the domestic politics of foreign policy making. Secondly, we will assess the operating experience with the Interim Arrangements between 1964 and 1972. Thirdly, there will be an analysis of the negotiations leading to the initialling of Agreements for Definitive Arrangements in 1971. And, lastly, we will address ourselves to the emerging opportunities and challenges associated with the innovation of distribution and direct broadcast satellites.

### Domestic Politics of Foreign Policy Making

After 1962, a three-pronged effort to develop operational space communications technology proceeded within the United States. One sees the programs of Comsat, the Defense Department, and NASA. It has already been established that there is a certain amount of political, economic, and technical intermeshing between these programs. But let us here elaborate the position of Comsat as the principal participant in implementing American communications satellite policy. How cohesive has the Corporation been? What purposes has it pursued? How have these purposes reflected the attitudes and actions of the President, the Director of Telecommunications Management, the Office of Telecommunications Policy, the FCC, and the State Department in light of their responsibilities for policy? What have been the characteristics of the oversight role of the Congress?

### Comsat's Position

From 1963 to 1969 the directors of Comsat were divided between six representatives of the public shareholders, six representatives of the carrier shareholders, and three Government appointees. The motive for this internal division of power was to insure that Comsat would not develop into an exploitative monopoly. There is an internal system of checks and balances as well as an external system, at least in theory.[a] Just how has this internal

[a]See p. 50.

137

competitive system reflected on the ability of the Corporation to act as the chosen instrument of American foreign policy? First, it should be pointed out that the Presidential appointees are not vehicles of Government influence on the board. Rather they are public directors responsible to the public shareholders' interest—not the Government's policy. This distinction and interpretation was made by the Attorney General in 1962.[1] While the Presidential appointees might represent what the Government considers to be the public interest in a certain situation, this is not preordained. The Government directors do not meet regularly with Government officials and do not attend every meeting of the board.[b] In addition, their credentials, while quite exemplary, do not equip them for arguing technicalities with representatives of the common carriers.[c] These factors tend to cast the Presidential directors in the role of odd man out, but how they have actually voted at board meetings is not public information. One might imagine, however, that they would side with the public directors against the carrier directors in any controversy, because the rationale of the Communications Satellite Act was to avoid carrier domination, especially AT&T domination, of space communications.

The carriers were entitled to six directors initially, and three represented AT&T, two ITT, and one HTC. However, by 1969, the common carriers, particularly ITT, RCA, and GT&E, had sold approximately 20 percent of their stock to the public. Congress amended the Communications Satellite Act to reflect these stock transactions, and the carriers are now entitled to only three representatives.[2] This diminution of the power of the carriers increases the chances that the conflict-of-interest potential built into the Corporation by the Communications Satellite Act will not be realized. It is thus increasingly the case that Comsat will press its own technology irrespective of the impact on the old technologies. However, the behavior of the carrier directors, all now representatives of AT&T, is still the key to the success or failure of the system of internal checks and balances. It is in the working relations between these directors and the public directors, who include the Chairman and Chief Executive Officer and President of the Corporation,[d] that one can gauge the health of management. Is this relationship characterized by anarchy or mutual adjustment? The answer depends on the issue under consideration, but my reading of the evidence suggests that Comsat decisions reflect a compromise between the interests of the

[b]Interview, Thomas E. Nelson, Foreign Service Officer in the Telecommunications Division of the Bureau of Economic Affairs, April 20, 1966.

[c]The Government directors are Frederic G. Donner, former Chairman of the Board of General Motors; Leonard Woodcock, President of the UAW and Rudolph A. Peterson, former President, Bank of America.

[d]The Chairman and Chief Executive Officer from 1965 to 1970 was James McCormack who replaced Mr. Welsh. Mr. McCormack is a retired Air Force General and was Vice President of the Massachusetts Institute of Technology. He is now chairman of the Board of Aerospace Corporation and was replaced as Chairman of the Comsat Board by Joseph H. McConnell, who is also President of the Reynolds Metals Company. The President since 1963 has been Joseph V. Charyk, a former Undersecretary of the Air Force.

carriers and the interests of the Corporation, conceived of as a competitor of the carriers.

The principal carrier, AT&T, has four roles with respect to Comsat: (1) as a stockholder—the largest, controlling 29 percent of the total stock; (2) as a customer—the most significant, using approximately 60 percent of all circuits leased by Comsat; (3) as a competitor with a stake in cables; (4) as a supplier through its subsidiary, Western Electric.[3] ITT had a similar set of functions in relation to Comsat. This disparity of roles could lead to fragmentation within Comsat, but so far it has not. The carrier directors recognize the potential for anarchy and abstain from voting on matters directly affecting their companies.[4]

There are three major issues involving relations between Comsat and the carriers, as well as other participants both inside and outside the Government. These are: (1) the earth station ownership issue, (2) the authorized users controversy, and (3) the issue of control and ownership of satellites for domestic communication. It is not surprising that the ramifications of these controversies involve foreign relations as well as domestic politics. But the major economic consequences of decisions on these issues principally affect the division of rewards between American corporations, and, thus, we will only treat them briefly here in relation to their interaction with foreign policy.

The politics of the earth station controversy sees the carriers aligned against Comsat. The carriers contend that the legislative history of the Communications Satellite Act provides for carrier ownership. Comsat argued in 1964 that, at least during the early stages, the Corporation should have undivided ownership and operational control. Dr. Charyk said that the principal reason for this was to avoid the multiple access problem. "A proliferation of ground stations will lead to a serious degradation of the capacity of the system which is undesirable, of course, both for technical reasons and economic reasons."[5] But the multiple access problem had been solved by 1966. Different ground stations can use the satellite at the same time without wasting frequencies or jamming each other. Nonetheless, Comsat still contends that it needs ownership of ground stations, not only in the early stages, but permanently. Why?

The principal reason for the earth station fight is probably the money involved. As earth station costs are considerable, there are large profits to be made from them, given the fixed rate of return allowed by the FCC.[6] But money is not the only motive for Comsat's position. Comsat argues that, if the United States is the only INTELSAT member without authority to control its own earth stations, this will inhibit Comsat's ability to demonstrate America's leadership role.[7] The carriers argue that Comsat should have the ability to control the earth stations as manager but that the carriers should own them, perhaps jointly with the Corporation.[8] Under an interim policy adopted in 1966 by the FCC, seven of the eight earth stations in the United States are owned jointly by Comsat and the other international common carriers. Comsat owns 50 percent of each, and the other 50 percent is divided among the other carriers.

The eighth station, in Alaska, is owned 100 percent by Comsat. However, in 1970, the FCC opened a review of its 1966 interim policy. The struggle thus continues with the FCC in the pivotal position.

The authorized users controversy, referred to in Chapter 6, was equally crucial to the financial position of the carriers, Comsat, and other entities, who are interested in dealing with Comsat without paying middleman costs to the carriers. Comsat wishes to be considered more than a carrier's carrier, but it does not want to go so far as to short-circuit the carriers' position. This would produce a conflict rather than a friendly competition, and Comsat realizes that the carriers will be customers as well as competitors. Thus, the Corporation has backed leasing circuits to users directly only if the carriers themselves fail to provide adequate service.[9]

Potential users of INTELSAT satellites such as International Business Machines, American Broadcasting Company, Columbia Broadcasting System, Associated Press, and United Press International have favored leasing directly from Comsat to cut costs. However, the carriers contend that Comsat was intended to be a carriers' carrier by the Communications Satellite Act. Authorizing Comsat to deal directly with users would, according to ITT, be devastating to the international common carriers. It "would allow Comsat to 'cream skim' the broad base of support which the total range of services offered by the existing record carriers rests upon; by siphoning off the largest users ... "[10] An anomalous part of this controversy is that some companies are both carriers and broadcasters. For instance, the National Broadcasting Company is part of RCA and wants direct access to Comsat, while RCA Global Communications argues for access through the carriers.

While principally a domestic issue, the authorized users decision also affected foreign policy, according to Comsat. Edwin J. Istvan argued that the United States could not be influential in pointing the way to other nations, if it could not put its own house in order.[e] But, while no one may doubt that Comsat's authority to deal directly with users other than the carriers would contribute to the cohesiveness of organization for space communications, one may wonder whether such a development would contribute to America's image as a free enterprise nation. Granting that Comsat is a monopoly established in the public interest, there appears to be no need to reinforce this situation.

The third major issue concerning the division of rewards from satellite communications technology involves the establishment of a domestic system. Is Comsat to be the chosen instrument of domestic policy as well as foreign policy? How will the answer to this question affect American foreign relations? The touchstone of this controversy was the 1965 ABC proposal to launch its own system to transmit television programs to stations affiliated with ABC-TV. Comsat contended that the ABC plan was contrary to the Communications Satellite Act. The Corporation saw its role as chosen in domestic as well as

[e]Interview, March 1, 1966.

foreign policy at least till March, 1971 when it proposed a joint system with AT&T, to the exclusion of noncarrier interests such as ABC, TRW, and Fairchild Hiller.

It is obvious that the prospects for revenue within the United States are greater than on the international scale, and, thus, Comsat's and other carriers' interest in a domestic system is quite natural. But how does this affect foreign relations? Some observers have expressed concern that the creation of a domestic system beyond the authority of INTELSAT would encourage the proliferation of other systems, which is contrary to the goal of establishing a single global system. On the other hand, including a domestic system within INTELSAT would assure other nations a source of income from what would be a domestic American system. For its part, the State Department has felt that a domestic system should not encourage other international systems but only other domestic systems, which is acceptable and desirable.[f]

In response to the ABC proposal, an FCC inquiry was undertaken in March, 1966 (Docket No. 16495). The Commission posed questions concerning its legal right to authorize a private, domestic system and the desirability of such a system in terms of economic, technical, and public interest considerations. However, so many conflicting views were expressed that it was not until June 16, 1972 that the FCC reached a decision on the matter. The details of this controversy do not concern us here for they almost entirely involve domestic politics. However, to lend perspective it should be pointed out that the financial stakes of the domestic and international issues are entirely out of proportion to each other. One must remember that AT&T has an investment of several hundred million in its international communications system, while it has $54 billion in assets in its domestic network. In 1966, the international carriers had income of $21 million, while the domestic carriers had income of $40 billion.[11] Of course, we cannot lose sight of the fact that some carriers such as ITT and Comsat derive most of their revenues on the international level.

*Executive Branch Activity*

Attention to making American foreign policy for satellite communications within the Executive branch, to a large degree, concerns relations with Comsat. As Comsat has been the manager of INTELSAT and the chosen instrument of American foreign policy, it is natural that this should be the case. In fact, while Comsat behaves primarily as a private enterprise, one must always bear in mind that it has two other very important roles: (1) as the quasi-Governmental representative of the United States in the international consortium and (2) as the manager and representative of an international organization, whose task is to build and operate the space segment of an operational, commercial communications satellite system.

[f]Interview, Thomas E. Nelson, April 20, 1966.

Government regulation of the foreign relations aspects of Comsat's activities involves not only the State Department but the FCC and the Office of Telecommunications Policy, which took on the functions of the Director of Telecommunications Management in 1970. In addition, NASA is intimately concerned, for under the Act it is responsible for aiding Comsat and giving technical advice to the FCC and the State Department.[12]

Between the negotiation of the Interim Arrangements, and the Definitive Arrangements, the State Department played a less active role in the policy process. According to Gilbert Carter, after State had helped set up the broad political parameters, it became Comsat's function to follow through with the day-to-day business.[g] This attitude corresponds to the idea that at times there is a distinction between foreign policy and business policy.

However, there is no indication that Comsat keeps to itself important information which might affect foreign policy. Before every bimonthly meeting of the Interim Communications Satellite Committee (ICSC), Comsat, as the American member, sends the proposed agenda to State for comments. And on crucial issues of policy, the State Department will issue instructions to Comsat as to the position it should take when acting as the American participant in INTELSAT.[13]

Within the Department of State, authority over relations with Comsat is centered in the Telecommunications Division of the Bureau of Economic Affairs. In January 1965, the President delegated his foreign affairs powers under Sections 201 (a) (4) and 201 (a) (5) to the Secretary of State, who in turn gave the Telecommunications Division the prime responsibility.[14] The main task this division has performed, since the establishment of INTELSAT, is to arrange for and persuade other nations to join the consortium. Whereas 19 governments initialed the Agreement in July, 1964, there were 83 by mid-1972. To a large extent this growth in the stature and size of INTELSAT has been due to the Telecommunications Division. This division was also intimately involved in the negotiation of Definitive Arrangements for INTELSAT.

The Federal Communications Commission is the most active Government agency in overseeing the performance of Comsat. It has the prime responsibility in such sensitive areas as earth station ownership, authorized users, and the establishment of domestic communications satellite systems. To the extent that these questions involve foreign relations, the FCC is part of the foreign policy process. Its mandate to reach decisions in line with "the public interest, convenience, and necessity" naturally involves a consideration of foreign policy and national security policy. The FCC is sensitive to this fact. One aspect of its decision on interim ground station ownership was a desire to strengthen Comsat's hand in the Interim Committee.[15]

In addition, the FCC has relinquished its control over the procurements in subcontracts of INTELSAT when the prime contract has been given to a foreign

---

[g]Interview, March 18, 1965.

company. This action lessened the concern of other nations that the FCC was regulating their affairs. The Commission was faced with a dilemma in this situation because, under the Act, it is supposed to regulate Comsat, yet Comsat is a member of an international organization as well as being a carrier and an instrument of national policy. The FCC cannot very well tell Comsat what to do if Comsat is acting as the legitimate representative of INTELSAT. It has been said that "our foreign partners are concerned that the FCC not become the WCC—the World Communications Commission."[h] Thus, the FCC must combine regulation of Comsat with promotion of American foreign policy—a delicate task so far successfully achieved.[16]

In its role as a promoter of United States foreign policy, it is obvious that the Commission is not acting as an independent, regulatory agency. It tries to coordinate with the Department of State. This is quite natural as what is involved is not adjudication of a domestic controversy, but promotion of the national interest abroad. There is some overlap here, however. A decision to strengthen Comsat in ownership of earth stations, for example, will run counter to the interests of the carriers and, perhaps, modify the intent of Congress and the antitrust laws. The FCC is usually described as an agency which protects the status quo, but in the case of the introduction of the new technology of satellite communications, the Commission must make a disruptive decision by the very nature of the case. Its role is to make and enforce rules assuring fairness in domestic corporate competition as well as to promote foreign policy.

NASA has a close relationship with Comsat, involving not regulation but technical coordination and cooperation. Under Section 201 (b) of the Act, it is clear that the Government is mandated to help Comsat as a fledgling enterprise in the same way the Government assisted the telegraph companies. However, the assistance in research, development, and launching is to be furnished on a reimbursable basis. There are to be no outright grants, but whether NASA's relations with the Corporation constitute a form of subsidy is a matter of political debate. The concern expressed during the 1962 debate and at the Incorporation Hearings is an indication of this.

If one takes the estimates on research and development programs by NASA, the Department of Defense, and Comsat (Table 8-1), one can see the extent to which the Government is still heavily involved in communications satellites. From this table one can see that NASA and the Department of Defense have spent more on R&D than Comsat. It is also interesting to note that AT&T had spent $76.4 million through 1964 on research projects relating directly to communications satellites, and is still heavily involved in such research through its interest in domestic systems.[i] These figures indicate how new an enterprise Comsat is and to what extent its technological base is built upon Government and AT&T expenditures.

---

[h]Interview with Asher Ende, Deputy Chief Common Carrier Bureau of FCC, May 2, 1966.

[i]Letter from George V. Cook, AT&T, August 17, 1965.

Table 8-1
**Government Estimates for Communications Satellite Research
and Development Program**[a]

|          | 1964 | 1965 | 1966 | 1967 | 1968 | 1969 | 1970 |
|----------|------|------|------|------|------|------|------|
| NASA     | 25.9 | 30.7 | 32.9 | 26.4 | 24.2 | 27.8 | 40.0 |
| Defense  | 80.2 | 25.7 | 53.9 | 62.3 | 18.3 | 63.2 | 73.4 |
| Comsat   | 8.7  | 21.3 | 39.5 | 62.4 | 4.0  | 7.2  | –    |

[a]In millions of dollars for Fiscal years 1964 through 1967 for NASA and Defense, information comes from material presented by the Director of Telecommunications Management to the Senate Committee on Aeronautical and Space Sciences, National Communications Satellite Hearings, p. 99. For Fiscal years 1964 through 1970, information comes from letter of R. Gould, Chief, Advanced Technology Division, Office of Telecommunications Management, Executive Office of the President, to Mr. Herbert Atkinson, Acting Chief Clerk, Senate Armed Services Committee, April 15, 1970.

Another indication of the Government subsidy being given to Comsat comes by way of a General Accounting Office (GAO) preliminary report which contends that the Air Force has undercharged launches for Comsat by $6.7 million through Fiscal year 1969. For example, on the first Early Bird shot in 1965, the Air Force costs were $922,110 but only $23,557 was billed to NASA, which in turn billed Comsat. The GAO report asserts that Air Force billing practices fail to conform to Federal statutes, Presidential guidelines, and provisions of the Communications Satellite Act (Sec. 201 (b) (3) ).[17] While the outside observer cannot determine whether a mistake or a conscious policy of subsidization was involved, it is a fact that the consequence of these events was advantageous to Comsat.

NASA, under the National Aeronautics and Space Act, is required to develop technology for the exploitation of the peaceful uses of space, communications being one of these. But NASA is an experimental, not an operational agency. Therefore, NASA concentrates its studies on developments that are many years away. Presently NASA is working on the Applications Technology Satellite (ATS) Program. Five spacecraft have been authorized under this program. The first was launched December 7, 1966. Each ATS experiment carries meteorological, communications, and other devices. According to NASA, its "R.&D. effort in communications experiments is designed to investigate the technical capabilities required for such future applications as navigation and traffic control satellites, distribution and direct broadcast satellites, and communications and navigations satellites to support deep space missions."[18]

The ATS program has evoked some concern in Congress as a long range study of Comsat's needs. For instance, Senator Clinton Anderson, Chairman of the Space Committee, asked a NASA official, "Are you really sure you are not doing things that the Communications Satellite Corp. should pay for?"[19] Regardless of whether NASA's ATS program would help Comsat or not, it might be needed in the Government's own interest. The difficulty with this policy lies in the dissemination of general knowledge gained from the ATS program. While it is acceptable for the Government to distribute knowledge gained from its own research and development activities, this distribution should benefit all companies in a given field, at least in theory. In the Comsat case, however, even the theory does not work, for the Corporation is a monopoly as far as international communications is concerned.

Comsat, on its own account, has grounds for worry about the ATS program. With a capacity of about 600 two-way voice circuits, an ATS could be used for commercial purposes or aid to less developed countries, after it completes its experimental program.[20] This could take away business from Comsat. Whether this contingency will ever occur, however, is questionable. In his 1965 report to the Congress on communications satellite activities, President Johnson made it crystal clear that the Government would only use its own systems in exceptionable circumstances.[21]

Until 1970, a central role in space communications policy had been given to the Director of Telecommunications Management. Under Executive Order 11191 of January 4, 1965, the DTM was assigned responsibility to generally advise and assist the President in connection with the President's exercise of his powers under Section 201 (a) of the Communications Satellite Act. In addition, the Director was to "serve as the chief point of liaison between the President and the Corporation."[j] But, in the late 1960s, there was much pressure from Congress and within the Executive branch to overhaul telecommunications policy and bureaucracy. Thus, on August 14, 1967 President Johnson appointed a Task Force on Communications Policy charging it to examine questions of frequency allocation, purposes, ownership and kinds of domestic systems, and the effect of these questions on Comsat and the international common carriers.[22] This task force headed by Eugene V. Rostow, Under Secretary of State for Political Affairs, finished its report on December 7, 1968, shortly before the Nixon Administration took office.[23] Like previous studies in the 1940s and 1950s,[k] this report, known as the Rostow Report, suggested numerous innovations to cure the hodge-podge of the nation's communications policy and organization, both foreign and domestic. The principal recommendations of the report concerned (1) integrating the various roles in which the Executive branch was currently engaged in a fragmented manner, i.e., setting up a Department of Communications;[24] (2) forming a single entity for international telecommunications, i.e., merging Comsat, AT&T, ITT Worldcom, RCA Global Communications, Western Union International, etc.;[25] (3) establishing a pilot domestic satellite program which would be managed by Comsat and would not be incompatible with INTELSAT in either technical or policy matters;[26] and (4) starting feasibility studies and modest pilot projects for introducing

[j]Interview, Herbert Roback, September 6, 1968.

[k]See Chapters 2 and 3.

telecommunications technology, especially of an educational variety, into the less developed countries.

How were the recommendations of the Rostow Report received by the Nixon Administration? In the first place, President Nixon has attempted to add greater direction to Executive Branch activities by centralizing responsibility in a new Office of Telecommunications Policy (OTP) in the Executive Office of the President.[27] Perhaps this new office will cure the communications hodge-podge and confusion in the Executive branch, but it seems that through mid-1972, the OTP has not really enabled the Government to speak with a more unified voice. No decision has been made to merge the international carriers into one entity; nor has there been a decision to continue with the present conglomeration of participants. In addition, the OTP was unable for over a year to get the FCC to move on domestic satellite communications policy, although in keeping with the ideology of free enterprise, the White House issued a very controversial policy statement in favor of competition between different domestic communications satellite systems, rather than a monopoly system as is essentially the case with telephone. The FCC, which, in any case, is somewhat independent of the Executive branch where domestic politics is concerned, finally decided in favor of a multiple entry, competitive policy on June 16, 1972, although the inquiry into this matter was initiated on March 2, 1966, and final comments were due on April 3, 1967.[28] Another area where OTP views have not been accepted as authoritative is in the case of the proposed Aerosat system for improving air traffic communications.[29] In fact, there are so many issue areas where the theory of centralizing power in OTP by Executive order has not been fulfilled in practice, that the trade press has referred to OTP Director, Clay Whitehead, as a czar without a czardom.[30] However, to balance this picture there are at least two areas where OTP has played a salutary coordinating role—in the latter stages of the negotiations to establish definitive arrangements for INTELSAT and in the preparation of the American negotiating position at the Second World Administrative Radio Conference for Space Telecommunications in Geneva, Switzerland in June and July, 1971,[31] Still, the total picture continues to look like the old pluralistic system which has been with us since the beginning of communications policy-making.

## The Oversight Role of Congress

Congress has generally played a low-key role in investigating developments in space communications from 1963 through 1971. The committees which have concerned themselves with policy and organization in this area are the House Military Operations Subcommittee, which was instrumental in overhauling the Defense Department's negotiations with Comsat and influenced the comprehensiveness of the decision-making process during the NASA negotiations with

Comsat for the NASCOM system; the two space committees, which have been primarily interested in overseeing NASA programs, the House Subcommittee on Space Science and Applications, which has issued far-reaching evaluations not only of international telecommunications but domestic satellite communications and organization in the Executive branch;[32] the Subcommittee on National Security Policy and Scientific Developments of the House Committee on Foreign Affairs, which has analyzed the consequences for foreign policy of the introduction of direct broadcast satellites;[33] and the Subcommittee on Communications of the Senate Commerce Committee, which has been principally concerned with the proposals for a domestic satellite communications system.[34]

From this brief listing, one can see that there is a fragmentation of jurisdiction in Congress where matters of space communications policy are concerned. However, this fragmentation of concern has not meant that each Committee or Subcommittee has restricted itself to analyses of problems within its purview. Since satellite communications developments affect a multitude of issues, every Committee has tended to touch on almost every issue. What has been lacking is an overall analysis of space communications in relation to the general and specific goals of American foreign policy. Policy making tends to be incrementalist, and so does the oversight process. One must refrain from saying that this is irrational merely because there has not been an explicit ordering of priorities and means to achieve them. It is unlikely that a person would know precisely what his goals were until he had muddled through a goodly number of concrete cases.[35] And, as Congress does not by and large, consider itself a coequal branch in the making of foreign policy, it is not unnatural to expect that its analyses of foreign relations problems will lack clarity and comprehensiveness until after the event. But this is also often true in the Executive branch, which does consider itself to be the central focus of foreign policy-making. What we are examining is not a case of Executive-Congressional relationships, but legislative politics according to the pluralist model where power is so dispersed within the Congress *and* the Executive branch—not to mention the FCC—that only a process of ad hoc coalition building and bargaining results in action.[36] Let us now evaluate the outcome of this fragmented policy making process, rather than the process itself, to see whether the nation's foreign policy has been adequately promoted.

## Assessment of INTELSAT, 1964-1972

The popular proposition that we live in an era when technological revolutions have left political man far behind is not true for communications satellites or other evolutionary technical innovations, as contrasted with revolutionary changes such as those in military explosives and delivery vehicles. The political and technological developments associated with INTELSAT since its establishment by nineteen governments in 1964 have paralleled each other in an orderly and harmonious manner. Membership had increased from nineteen to seventy-nine by the time the Definitive Arrangements were initialed on May 21, 1971. Meanwhile, a single, global system serving the Atlantic, Pacific, and Indian Ocean basins had become technically operational in the summer of 1969. This system was based on the INTELSAT I, II and III satellites which provide 240, 240, and 1,200 voice circuits respectively, and in 1971, INTELSAT IV satellites were introduced with 3,000-9,000 circuits.

Not only has membership in INTELSAT increased as its technological promises have been realized, but, also, the costs of communications by satellite have decreased. For instance, initial costs to American carriers of $32,000 per half-circuit per year were reduced to $20,000 in 1966. In addition, the integration of satellite communications into the existing international communications structure savings to the Government alone amounting to $9.5 million annually.[37]

There are political and technological reasons for the evolutionary pace of change in space communications. Politically, at the domestic level in the United States, the Federal Communications Commission and dominant traditional interest groups, mainly the common carriers, have worked to control innovation in order to protect investments in traditional microwave and cable facilities. One example of this is the April 1972 FCC decision to allow AT&T to lay a new cable in 1976 which may be less efficient than an INTELSAT IV satellite, the first of which was launched in 1971. The new cable will have a capacity of 4,000 circuits whereas INTELSAT IV's capacity is between 3,000 and 9,000 circuits. The cable will cost over $100 million, while the satellite costs $13.5 million. But one must consider associated costs to arrive at a true comparison, and this is a very difficult matter. At a minimum, one has to calculate the cost of the launch and at least two earth stations into the true cost of the INTELSAT IV circuits. This would add about $16 million for launch and $6 million for two earth stations or a total of $35.5 million for each satellite.[38] This is still a third of the cost of the new cable. To justify the cable, one must then demonstrate that cables are more reliable than satellites due to such factors as earth station outages. Perhaps this can be done, but it would take some convincing given the $65 million involved. Another indicator of slowing down innovation is that, despite the continental size of the United States, it is likely that domestic space communications systems will be operational in Canada and India before the United States. The Soviet Union has the only domestic system at present—ORBITA. Another political and economic reason for incremental change is the attitude of certain foreign governments. The United Kingdom, like AT&T, has a vested interest in depreciating its cables before backing space communications inventions on grounds of efficiency.

In addition, given the persistence of nationalism, certain countries have expressed desires to establish regional systems rather than channel all traffic through INTELSAT. The French-German SYMPHONIE project is an example. This policy works to slow down change in INTELSAT because these nations have not yet developed the technical capability to build and launch comsats on an economically profitable basis.

Technological factors have also made communications by satellites evolutionary rather than revolutionary. The early uncertainty concerning the choices between random and synchronous satellites, which was eventually resolved by the developments associated with the Early Bird and Syncom satellites, worked at the time to make change more piecemeal. Once this hurdle had been crossed, other technical issues arose. Multiple access to satellites from different earth stations was achieved. The ability to focus the signals of satellites on particular areas of the earth—the narrow beam capability—was achieved, thus further refining the state of the art by increasing the effective power of satellite signals, which consequently reduces earth station costs. Additional technical developments of importance to the efficiency of space communications are the ability to move satellites in orbit from one slot to another and to decrease the distance between one satellite and another in synchronous orbit without increasing noise.

All these advances in the art have been refinements and evolutionary changes which have increased point-to-point communications capabilities. Perhaps other innovations, e.g., direct broadcast satellites, could have immediate revolutionary impacts on world communications patterns; but, at this time, changes have produced a series of incremental improvements.

There are technical developments on the horizon of a competing nature which could conceivably undermine the economic promise of satellite communications. AT&T plans a short-haul cable in 1973-74 with 105,000 circuits, and, by the late 1970s, AT&T foresees transmissions by waveguides with 250,000 circuit capacities. These may be followed by laser tubes with up to two million circuit capacities.[1] Such developments would surely call into question the economic potential of satellites with their need for increased power sources and expensive ground equipment. To the extent that competition from other techniques threatens space communications, there exists an additional limit on a rapid and thorough reliance on space communication for long-haul operations.

Having established the proposition that there have been evolutionary political, economic, and technological developments within the context of INTELSAT, let us now turn to a more detailed analysis of the experience with the Interim Arrangements and the issues which have been encountered in the transition between the Interim and Definitive Arrangements.

INTELSAT's principal purpose is to provide for the design, development, construction, establishment, maintenance, and operation of the *space segment* of the global *commercial* communications satellite system. It is important to realize, however, that a viable communications satellite system also requires earth stations, satellite launchers, and communications infrastructures in differ-

[1]Interview, Stephen Doyle, April 20, 1972.

ent countries. Nations own their own earth stations, while the satellites are owned internationally. Only the United States, the Soviet Union, France, Japan, and China can launch earth satellites, and the communications facilities of many countries have exceedingly sparse numbers of telephones, TVs, and radios. Another basic distinction is between INTELSAT as a commercial organization and one serving broad political goals. Although the Agreement describes INTELSAT as a commercial organization, many of its members view it as a global public utility,[m] and the Preamble states that governments desire "to establish a single global commercial communications satellite system as part of an improved global communications network which will provide expanded telecommunications services to all areas of the world and which will contribute to world peace and understanding." This Preamble resembles the Communications Satellite Act, which is even more specific in its language relating to broad political objectives. It is thus appropriate to analyze INTELSAT's political and social, as well as economic, functions and structures.

A first approach to such an anlysis is to explicitly relate INTELSAT's functions, defined as goals, to its structure. I have constructed a structural-functional matrix in which + means that in my judgment the relation between structure and function is positive; − means the relation is negative; and 0 means there is a nonfunctional relationship.[39]

If the signs in the matrix are correct, the number of plusses indicates a considerable meshing of structures and functions; under the Interim Arrangements structures of power overlapped to a high degree because Comsat was not only manager but owner of over 50 percent of INTELSAT's space segment, and this is its voting share in the joint venture as well. INTELSAT could still be called a "supranational" partnership, however, for, in spite of Comsat's dominance, the Interim Arrangements have (1) weighted voting rules where Comsat may be vetoed, and the absence of the one-state—one-vote rule; (2) binding arbitration in case of legal disputes; and (3) devotion to the attainment of broad political as well as commercial goals. Whether this integration of the satellite communications agencies of 83 nations into one international organization will continue under the Definitive Arrangements depends on the dialectic, as Galtung describes it "What is forgotten is the changing nature of these relations over time when plusses are increasingly taken for granted, and frustrations due to minuses accumulate."[40] Therefore, even if the evaluation of integrative accomplishments is correct, no basis for future understanding or prediction is provided. A more differentiated structural-functional matrix will characterize INTELSAT in coming years.

---

[m]If INTELSAT is analogous to a public utility, a comparative analysis might well see it in relation to international public unions such as the Universal Postal Union (UPU), or, at least, organizations such as the International Air Transport Association (IATA), which has been quite successful in the field of commercial aviation. However, these associations are more like regulatory agencies. They do not own post offices and airplanes, as INTELSAT owns the space communications satellites.

**Table 8-2**
**Structural-Functional Relationships in INTELSAT**

| | $f_1$ | $f_2$ | $f_3$ | $f_4$ | $f_5$ | $f_6$ | |
|---|---|---|---|---|---|---|---|
| | Expand Capacity | Decrease Costs | Increase Membership | Distribute Contracts | Expand Commercial Service | Expand Educational Service | Promote Peace and Understanding |
| $s_1$ Voting and Ownership | 0 | + | + | + | + | 0 | 0 |
| $s_2$ Manager (Comsat) | + | + | + | 0 | + | – | 0 |

However, to help us visualize the global success of the INTELSAT system in mid-1972, Figure 8-1 shows the locations of INTELSATS and earth stations. Going beyond this we can supplement the matrix in Table 8-2 with a more detailed list of qualitative and quantitative measures of integration within INTELSAT and the integrative effects of INTELSAT on world communications patterns and institutions. These measures will be placed in political, economic and social categories.

### Political Measures

1. *membership*—from 19 states in 1964 to 83 in mid-1972, including 1 Communist state, Yugoslavia. (see Appendix B).
2. *voting formula*—a weighted voting formula where Comsat may be vetoed, but in practice attempts are made to reach unanimous decisions. Of 300 decisions in the first five years, only 3 or 4 have been by less than 70 percent.[41]
3. *arbitration*—a binding arbitration agreement in case of legal disputes.[42]
4. *internationalization of management personnel*—as of December, 1969, 90 of Comsat's 1,005 employees were nationals of foreign countries, while 17 out of a professional staff of 641 were nationals of foreign countries.[n]

### Economic Measures

1. *ownership quotas*—as of December 31, 1970, the U.S. share was 52.61 percent, U.K. 7.25 percent, France 5.26 percent, Germany 5.26 percent, Japan 1.73 percent, etc.[43]
2. *market growth*—use of overseas channels grows at between 20-25 percent per year from U.S., and supply rather than

[n]Telephone conversation with Dr. Renfield (INTELSAT Affairs Division, Comsat), December 15, 1969.

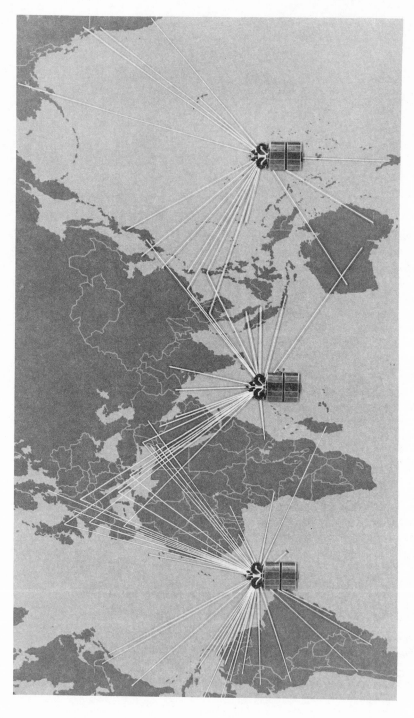

**Figure 8–1.** A global system of high capacity INTELSAT IV satellites in synchronous orbit over the Atlantic, Pacific and Indian oceans provide direct lines of communication among earth stations in more than 40 countries of the world. Satellites today account for more than two thirds of all overseas long distance telephone calls. In addition, one out of every six people on earth can see historic events as they happen on their TV screens via satellite. Source: Communications Satellite Corporation, Washington, D.C.

demand has limited growth in the past.[44]

3. *distribution of production*—subcontracts to non-U.S. firms in other countries have increased from 2 percent of contracts in the case of INTELSAT II to 6 percent for INTELSAT III and 27 percent for INTELSAT IV. In addition, 13 percent of research and development contracts are awarded outside the U.S. and this percentage is increasing.[45]

4. *channel capacity*—the number of full-time leased two-way circuits increased from 75 in 1965 to 2129 at the end of 1970. Global TV service had also increased from 40 half-channel hours in 1965 to approximately 997 in 1970.[46]

5. *access*—multiple access to satellites from earth stations. At the end of 1970, there were 43 earth stations in 30 countries; by the end of 1974, it is estimated that there will be more than 90 antennas in operation in about 58 countries.[47]

6. *relative cost*—the new transatlantic cable authorized by the FCC in April 1972 for service by 1976 will have a 4,000 voice circuit capacity and cost over $100 million for a planned life of 20 years, while the INTELSAT IV has a 3-9,000 circuit capacity at a cost of $35.5 million for a planned life of 3-6 years.[48]

7. *assets and capital expenditures*—the net book value of INTELSAT as of October 31, 1968, was $75,417,311, and in the years 1964 through 1970, the cumulative gross expenditures were $271 million. (The U.S. share was $142 million or 52 percent.)[49]

### Social Measures

1. *types of communications*—to date INTELSAT has handled TV as well as telephone and telegraph traffic. The TV capability of space communications satellites is an important innovation, the integrative effects of which are subject to diverse interpretations.

2. *content of communications*—all overseas carriers handle predominantly commercial and governmental traffic. There are no plans to have INTELSAT serve mass educational or propaganda purposes.

The listing of measures for the political, economic, and social integrative aspects of INTELSAT along with brief explanations of each measure is evidence for the proposition that INTELSAT is characterized by a high degree of political and economic, if not social, cohesion.

Another question to ask concerning the integrative aspects of INTELSAT is whether there has been any external spillover from this enterprise to other endeavors. How has INTELSAT affected the integration and rationalization of the world's communications patterns and institutions? One may at least say that it has increased international communications capacity, lowered the cost of international communications, and worked to overcome regional and neocolonial communications ties with a single, global, multiple-access system. There has been no effect of the cooperation in INTELSAT on integrating the world's cable and microwave relay structure. Also, INTELSAT integrates nations' foreign satellite communications traffic—not domestic satellite communications traffic (except in circumstances as when one part of a state is separated from another by geographic barriers or the jurisdiction of another state, e.g., Continental US and Hawaii or Alaska). Granting increased economic integration has occurred because of INTELSAT and satellite communications technology, however, one may go on to ask what has been the importance of these changes?

In 1966 the International Telecommunications Union pointed to the promise of modern telecommunications as an instrument "to transform the nations of the world into a single community." But if modern technologies of communications are extensions of man's senses, to use Marshall McLuhan's terminology, then who is to say that man will use his new tools of communication not to understand his fellow man but to massage him? Technically, we may be more and more one world in that it takes less time to send messages, but socially we may be growing apart or standing still. It is not only the potentialities of technology that form criteria for evaluating integration and interdependence. An equally if not more important criteria is the *content* of the messages which the new means of communication send—to dispute McLuhan. Further, it is not only the content but how people interpret the content of the messages they receive. The most we can say about technological causation is that satellite technology presents the *capability* for more political, economic, and cultural integration.

Under the Interim Arrangements, there has been an increase in membership, markets, and the number of circuits, and a decrease in costs. Perhaps the one failure in INTELSAT's stated goals is that of establishing a *single* worldwide system. The Soviet Union and its allies have not joined but have instead indicated that they are interested in their own international space communications system, INTERSPUTNIK. But by 1972 INTELSAT remained the only operational international space communications venture. And the fact that the Russians have not joined does not mean that they will not cooperate with INTELSAT.[50] For instance, President Nixon's summit meetings in Moscow (and China) were covered through arrangements for TV links with INTELSAT. Even if the Soviets do not cooperate, however, the loss in terms of integrating the existing international communications flow will not be great. Members of INTELSAT now account for more than 95 percent of international traffic. Thus, INTELSAT is obviously not a regional, but a global organization.

In conclusion, an evaluation of the experience with INTELSAT to date points to its positive if not dramatic effects on the integration of the communications networks of eighty-three different states. This has been both an economic and political accomplishment if not an example of social or cultural interdependence between nations.

**Negotiation of the Definitive Arrangements**

Article IX of the Interim Arrangements called for the negotiation of Definitive Arrangements for INTELSAT in 1969. In preparing a negotiating position for the Conference, a policy group was established consisting of representatives of State, FCC, DTM, and Comsat. From time to time representatives of NASA and the Department of Defense would also attend. This policy group met weekly or biweekly during most of 1968. There were no deep divisions within the group, but there were opposing views expressed concerning the place of the manager and the permissibility of regional systems under permanent arrangements.[o] In the first case, Comsat thought, in contrast to the government representatives, that it should continue to be the sole manager. The Corporation sought to muster strength in its favor through appeals to supporters in Congress such as Senator John O. Pastore (D., R.I.) and Representative Harley O. Staggers (D., W.Va.). But the Executive branch was supported by other international carriers such as AT&T and RCA.[51] In addition, the Twentieth Century Fund issued a report backing proposals which would have undermined American control in INTELSAT in a much more drastic manner than the State Department or the FCC envisaged.[52] As a result of the bargaining between these various interests, Comsat was to become tacitly reconciled to a solution which would divide the managerial functions in two parts—administrative/financial as distinguished from technical/operational. Comsat would retain management functions in the latter areas while the former would be internationalized.[53] On the second issue, Comsat and DTM did not want any regional systems to be permitted under the Definitive Arrangements, but the final, unified position was in favor of a single system while not ruling out regional systems if they were technically and economically compatible with INTELSAT.

From this brief review of the work of the policy group, one might wonder if America had compromised the short-range exclusive economic interests of Comsat in favor of the long-range inclusive interests of the United States and the world. There is no evidence to support this conclusion for, during the negotiations in 1970, State defended INTELSAT to the Congress in terms of narrow economic interests. At the end of 1970, Mr. Samuel de Palma, Assistant Secretary for International Organization Affairs, and Ambassador Abbott Washburn, Chairman of the American delegation to the INTELSAT Conference, presented a request to a Subcommittee of the House Committee on Appropriations for additional appropriations to continue the Conference. In this request, it was argued that, in 1964, the United States had been able "to negotiate a controlling position (in INTELSAT) . . . because we negotiated from strength."[54] However, in the intervening six years "contentious" problems had developed. Perhaps in order to get $218,000 from Representative John J. Rooney's (D., N.Y.) subcommittee, State defended INTELSAT from the point

[o]Interview, Stephen E. Doyle, Foreign Affairs Officer, Office of Telecommunications, Department of State, July 23, 1969.

of view of narrow self interest, particularly economic interest. This excerpt from the statement is indicative:

Since 1964 the 76 members of the INTELSAT consortium have invested $350,500,000 in the system and America's share (and voting power) is currently about 52% or $266 million. Ninety-two percent of the total spent ($323,500,000) went to American contractors.

In addition, the various members of the consortium have built and already have in operation some 50 earth stations in 28 countries involving an average investment of $5 million per earth station. This represents a total investment of $250 million. Our conservative estimate is that American manufacturers have provided in excess of 50% of the hardware in these earth stations for a total of more than $125 million in actual business.

INTELSAT was not only defended as a source of contracts to American business at home, but also abroad:

The INTELSAT system has also brought modern and direct communication to many areas of the world which previously had none, thus enabling American business to better utilize its operations in many countries.[55]

It seems reasonable to conclude from this testimony that America was not neglecting its short-term interests, although it is possible to argue, as this author does, that in the drive to expand communications facilities around the world, everyone can benefit.

In addition to the domestic preparations for the 1969 negotiations, there were many contacts between Comsat, the State Department, foreign governments, and telecommunications agencies. Some of these contacts were institutionalized through the Interim Communications Satellite Committee, which was responsible for rendering a report on the Definitive Arrangements. Other contacts were informal. What one sees even more clearly than in 1963, is a decision-making process which cut across national boundaries. The process was pluralistic and legislative rather than elitist and hierarchical. The politics was low-key and conducive to bargaining and coalition-building between and within states. Neither the United States, the Western European states, nor the less-developed countries saw their vital national interests at stake.

There was no cohesive European view on policy for the Definitive Arrangements.[56] One might have expected that the European Conference on Satellite Communications, organized in 1963 and better known as CETS after its French title,[P] would have been able after five years of experience to contribute to a unified European position, but CETS was more a forum for discussion than a spokesman for Europe. One reason for this was that underindustrialized states such as Spain and Portugal did not have the same economic motives to overcome the "technology gap" as did France or the United Kingdom. Furthermore, the stake of the United Kingdom in the old terrestrial facilities was greater than that

[P]See Chapter 5.

of Germany, for example, and thus the highly industrialized European States did not necessarily see eye to eye. In general though, there was a strong feeling that American power in INTELSAT should be lessened. Specifically, many countries wanted to remove Comsat as manager, restrict its voting power, and undercut INTELSAT through denying the single global system goal. This last point could have formed the basis of a more bitter conflict, if the various European space organizations had been able to build launching rockets for their satellites. CETS had proposed a point-to-point satellite for television service to become operational in 1970, but the ELDO rocket launcher for this program has never materialized. In 1968 the British had announced that they were going to withdraw from ELDO in 1971, and by 1970 it had become clear that the hopes of the Europeans for their own regional communications satellite system depended upon American launching rockets.[57] Another regional program which has not materialized is the SYMPHONIE project organized in 1967 by the French and the Germans. The French are the only Western European nation to have launched a satellite of their own, and therefore this project could succeed. On the other hand, the lifting capability of French rockets is restricted, and it seems much more economical to many Europeans to buy rockets from NASA or the Soviet Union than to start from scratch. In fact, in spite of European complaints of American dominance in INTELSAT, all European countries combined spend only $300 million a year on space research, less than one-tenth the expenditures of the United States or the USSR.[58] Thus, cooperation with INTELSAT seemed the better part of wisdom to most Europeans.

The less developed countries, for their part, were quite anxious to continue with INTELSAT, which offered them capabilities at costs which were far below those of traditional technologies. They did, however, want to increase their power within INTELSAT, perhaps by substituting one state, one vote for the weighted voting arrangements in the ICSC. In addition, some delegations from the less developed countries wished to turn INTELSAT into an international organization like the UN, with a Secretary-General and a General Assembly to decide all policy, but minus a Security Council with the veto for the major states. Such a view was anathema to all opinion within the American negotiating team.

It was natural that the divisions within states and between them did not give rise to a cohesive report by the ICSC. There was some unanimity on certain points, in particular; that INTELSAT should be a world organization devoted to global telecommunications services; there should be nondiscriminatory access to the system; INTELSAT should only own the space segment of an international communications satellite complex; and there should be two agreements as in 1964, one between states and the other between designated telecommunications entities. The report indicates that the following principal issues arose to be settled by the Plenipotentiary Conference.[59]

1. To what extent should INTELSAT's structure, voting, and ownership arrangements be changed to increase the equality of states and internationalization of the consortium?
2. To what extent should the management function be internationalized?
3. Should there be a single global system or several international and regional systems?
4. Should INTELSAT handle specialized services, e.g., mobile and direct broadcast, or just traditional point-to-point communications?
5. Should INTELSAT continue as a joint venture or be given legal personality?
6. Should INTELSAT issue procurement contracts on a purely competitive basis?

These major areas of controversy reflected, in essence, an attempt to revise the status quo within INTELSAT towards a more egalitarian structure, where Comsat's financial, technical, and political influence would be lessened. In addition, there was not only a focus on limiting the American role in INTELSAT but limiting INTELSAT's role vis-a-vis other possible international communications satellite systems and services. These major areas of competition also characterized the negotiation of the Interim Arrangements in 1963-1964. However, competition must not be confused with outright conflict. The principal issues were disagreements among partners rather than fights between enemies. It is interesting to note that the major issues of 1963-1964, i.e., allocation of ownership quotas and voting arrangements (with the obvious exception of how long the arrangements would last) were also principal issues in the Definitive Arrangements negotiations. It is true that there was a change in the specifics of each issue and also there was proliferation of new issues, but this was because the participants had increased from nineteen to seventy-nine.

The Plenipotentiary Conference convened on February 24, 1969, in Washington, D.C. and after intensive, long, and delicate negotiations, the Definitive Arrangements were initialed on May 21, 1971, after having been approved by a vote of 73-0, with 4 abstentions and 2 absent. The two agreements were opened for signature on August 20, 1971. They are called the Agreement and the Operating Agreement Relating to the International Telecommunications Satellite Organization. These Definitive Arrangements will enter into force 60 days after the intergovernmental agreement has been adhered to by two-thirds (54) of the 80 governments which were parties to the Interim Agreement as of August 20, 1971.[q] Let us now analyze each of the principal issues in order to clarify the varying viewpoints and the solutions which were finally developed. Rather than develop the analysis chronologically over the 27 months of negotiations, we will focus on each issue and its solution.[60]

[q]The Definitive Arrangements had not entered into force by mid-1972. A copy of the Arrangements may be found in Senate Aeronautical and Space Sciences Committee, *International Cooperation in Outerspace Symposium*, Appendices 1 and 2.

*Structure, Voting, and Ownership*

Under the Interim Arrangements, as discussed above, INTELSAT basically had two structures—the manager (Comsat) and the ICSC. It was apparent to all parties that the Definitive Arrangements would be more differentiated in order to reflect the principle of equality, i.e., where states would vote on a one state, one vote basis. It was possible that the Definitive Arrangements could have had five structures: (1) a conference of states, (2) an assembly of operating telecommunications entities, (3) a board of governors, (4) an international administrative manager, and (5) a technical manager which would be Comsat for several years. In the final agreements, provision is made for a four-tiered arrangement composed of (1) the Assembly of Parties, (2) the Meeting of Signatories, (3) the Board of Governors, and (4) an executive body, responsible to the Board of Governors. However, the basic issue was not the number of structures but the relative power of the structures in relation to each other.[r]

The less developed countries generally wanted a large role for the Assembly of Parties, which, being governed by a one state, one vote formula, would give them the real power in INTELSAT. The United States and most developed countries favored having control lodged in the Board of Governors where weighted voting would be the rule.[s] During 1970, Britain urged the United States to give formal powers to the Assembly, which would meet only at two year intervals, thus assuring that real control would reside in the Board of Governors. But the United States refused this suggestion, asserting that the Assembly should *recommend* rather than *decide* policy. In the end, the American position was substantially approved as the Agreement limits the Assembly to consideration of general policy with the authority to make recommendations to other bodies. Only in the case of the provision of specialized services is the Assembly given specific power to approve their inclusion through a two-thirds vote. In addition, the Meeting of Signatories, which is composed of operating telecommunications entities and is also governed by a one representative, one vote basis, has only powers of recommendation with certain exceptions relating to the establishment of rules for the approval of earth stations, the allotment of space segment capacity, and the establishment of rates for the use of the space segment. These powers, which are real enough, will in practice be less extensive than it might seem as the Meeting of Signatories will normally meet only once a year and will make its decisions upon the recommendation of the Board of Governors. Thus effective power resides with the Board of Governors, whose functions include the design, development, construction, establishment, operation, and maintenance of the INTELSAT space segment. This solution parallels the Interim Arrangements where effective control resided in the ICSC. This reflects the principle that control of INTELSAT should be weighted to favor those who use it most.[61]

[r]These issues can be analytically separated but are here linked together because of their close relationship in practice.
[s]Interview, Stephen E. Doyle, August 21, 1970.

This leads one into an analysis of the ownership and voting issues. There was general agreement that there should be a relationship between investment shares and voting rights on the Board of Governors, and that these in turn are related to use.[62] But, as in 1964, there was no agreement on the specific formula which would govern this relationship. There was agreement that the Board should have about twenty members. The ICSC had eighteen members and it seemed appropriate to keep the membership at about this level in order to guarantee that the Board would not become too unwieldy. Furthermore, it became generally accepted that no one member of the Board should have more than 40 percent of the vote even if its use of INTELSAT might entitle it to a greater amount. Thus Comsat's majority power was eliminated. A disagreement developed, however, concerning whether Comsat should retain a veto power. Under normal circumstances, substantive matters will be decided by a two-thirds vote thus assuring Comsat a veto. In the 1969 negotiations, the United States took a position in favor of retaining this veto in the face of growing resentment from abroad.[63] But in 1970, the United States gave in, and the Agreement establishes a formula whereby no three members, regardless of the size of their weighted vote, may veto an action supported by all the other members of the Board. The significance of this dispute is not great, historically speaking, as under the Interim Arrangements ICSC records show that most actions have been determined by consensus.

Another delicate matter concerned the provisions by which investment in the system are to be determined. In 1969, the United States insisted that share of traffic was the only rational criteria for determining investment, as was the case under the Interim Arrangements. Other proposals would have added additional criteria such as population and investment in earth stations of the global communications satellite network. These latter suggestions were proposed by some of the less developed countries that wished to increase their power on the Board of Governors; but in 1970, it was agreed that voting participation should be derived solely from the utilization of the INTELSAT space segment. However, use of the space segment is not restricted to international public telecommunications services, but also includes domestic services between areas separated from each other by other jurisdictions or exceptional natural barriers. Thus communications between the continental United States, Alaska, and Hawaii are included as are communications between the islands of Indonesia, Britain and Hong Kong, etc.

*Manager*

During the 1969 negotiations on the Definitive Arrangements, three views arose with respect to the manager issue. One group of Belgium, France, India,

Switzerland, United Kingdom, Canada, and Germany, proposed that there be a full internationalization of the management under a director general, within a specific period of time.[64] Secondly, a group composed of Australia, Chile, Nigeria, and Venezuela favored the first approach subject to the objection that they felt a fixed time limit for achieving internationalization might interfere with the objective of ensuring efficient and effective management. A third view, that of the United States, was that internationalization of the manager should not, in itself, be a primary goal or common aim of INTELSAT. Rather the United States felt that "efficient management should be the only goal of the organization regarding the structure of the management body. Internationalization of the organization should be addressed in the assembly and the governing body."[65]

How were these views reconciled? In March, 1970, Australia and Japan introduced a package proposal,[66] the major portion of which was devoted to the management issue. As mentioned above, the United States modified its opposition to internationalization of the management function by agreeing to the division of the function in two—an administrative, financial, and legal manager on the one hand and a technical, operational manager on the other. The first could be internationalized but the second would require proven expertise and would be lodged in Comsat under a management contract to the Board of Governors for a period of six years. In the Definitive Arrangements, it is stated that there will be a Secretary-General, who reports, but does not control, Comsat's activities to the Board of Governors. However, by not later than December 31, 1976, a Director General shall be appointed, and he will be responsible for *all* management services, both administrative and technical. In other words, by 1977, INTELSAT will be run by an international secretariat which is called the executive body. The Director General will head this body, and he will be appointed by the Board of Governors, subject to confirmation by the Assembly.

*Separate Regional or Global Systems*

Given the persistence of nationalism in the world, certain countries may seek to establish separate regional or global systems competitive with INTELSAT, e.g., the SYMPHONIE and INTERSPUTNIK proposals. The United States has been opposed to separate systems for many years on the grounds that they would economically undermine INTELSAT and increase the problems of frequency assignments, noise, and the distribution of orbital slots. Also involved has been the image of American leadership in a single global system. However, in the fall of 1969, the United States agreed to the establishment of regional systems if a two-thirds vote of the Assembly decided there would be no economic or technical incompatibility between the two systems. This major concession was necessitated by considerable pressure from a coalition of developed and less developed states.[67] In addition, the Soviet Union had sent an observer to the 1969 negotiations and he had expressed the view that there should be as many domestic, regional, and global systems as required to meet the needs of nations. By the time the final Agreement was signed in 1971, the door was left even further open to the possibility of competing systems, for any party has the right to set up a separate system, being restricted only by the need to inform INTELSAT of its plans and to seek to ensure technical compatibility with and avoid significant economic harm to INTELSAT. The Assembly cannot stop the establishment of separate systems but only recommend against their establishment. On the other hand, all members of INTELSAT have committed themselves in the Preamble to the Definitive Arrangements to a single *global* system. Consequently, the door is left open only for separate *regional* systems.

## Specialized Services

This controversy concerns whether INTELSAT should be permitted to provide specialized telecommunications satellite services, e.g., radio navigation, direct broadcasting, space research, meteorological and earth resource services, or whether the organization should be restricted to providing public telecommunications services such as were provided under the Interim Arrangements, i.e., telephony, telegraphy, telex, facsimile, data transmission, and point-to-point broadcasting. There was no disagreement concerning the provision of military communications as a specialized service as all delegations wished to exclude this traffic.

The Third Plenipotentiary Conference, meeting in April and May of 1971, was deadlocked on the issue of which, if any, specialized services should be provided by INTELSAT. The United States refused to support a majority position which had excluded mobile public telecommunications, i.e., ship-to-ship, ship-to-shore, aircraft-to-aircraft, and aircraft-to-shore, from INTELSAT's normal traffic. The leader of the majority was the United Kingdom which has considerable investment in mobile services using traditional microwave facilities; and the United States contended that this position would "benefit one or a few countries now enjoying a virtual monopoly in certain public telecommunications services via inefficient terrestrial facilities."[68] The United Kingdom, on the other hand, maintained that United States reasoning was faulty as the majority position "in no way bars Intelsat from taking advantage of technological breakthroughs. It simply makes it clear where the power of decision lies on whether Intelsat should expand its activities into certain new fields of a specialized nature."[69] That is, the United Kingdom position did not prohibit INTELSAT from providing specialized services; it only stated that this was a substantive matter to be approved by a two-thirds vote of the Assembly of Parties. Reluctantly, the United States acceded to this position.

It is not clear from the solution to the controversy whether specialized services will be introduced in a timely and efficient manner. The outcome will depend on the parliamentary situation within the Assembly of Parties. One aspect of this situation does deserve particular attention, and this is the timing and the context surrounding the introduction of direct broadcast satellites. Under the Definitive Arrangements, the Assembly can, by a two-thirds vote, approve INTELSAT as the appropriate organization to handle this potentially revolutionary technology. Significantly, however, the debate during the negotiations did not concern direct broadcast satellites but mobile services. The explanation for this lay in the parties' knowledge that this innovation could not be introduced until the late 1970s and thus the issues surrounding this technology were not thoroughly considered.

*Legal Status*

Another issue concerned the legal status of INTELSAT. The United States wanted the consortium to continue as a joint venture without legal personality. But the majority view held that INTELSAT would be better able to make contracts, own property, sue or be sued, obtain privileges and immunities, and incur and dispose of liabilities if it were a separate legal entity. Again, the United States gave in and accepted this view. This shift of position indicated that two problems relating to Comsat operations had been resolved. Previously, the Corporation treated its investment in INTELSAT as if it were a partnership, thus allowing for depreciation for tax purposes. But changing INTELSAT into a legal entity would not enable the Internal Revenue Service to continue treating Comsat's investments in this manner. To solve this problem, a waiver was granted in this particular case. Secondly, American companies have not been able to count their investments in multinational corporations as part of their rate base on which profit allowances are made by the FCC. Again, this practice has been breached in order to promote compromise internationally while maintaining the financial strength of Comsat.[70]

*Procurement*

As the 1968 ICSC Report on Definitive Arrangements points out, there are two considerations in regards to procurement which tend to be contradictory. "On the one hand, there is the need to obtain the best product at the best price. On the other hand, there is the legitimate desire of member states to share in procurements and to further their technological development."[71] The division of states on the weighing of these values saw the United States and most less developed countries aligned against Europe, Canada, and Japan. The unindustrialized states did not want to see an equal or proportionate-ownership procurement formula adopted for this would increase costs to them through subsidizing less efficient technology.[72] The United States favored a simple statement that contracts were to be awarded to the best bid. This would assure American business of the flow of most INTELSAT contracts for the foreseeable future. Europe, Canada, and Japan, on the other hand, wanted an additional proviso that in cases of equal bids, the deciding factor would be the extent of subcontracting among member nations. Again, the United States gave in, for the Agreement encourages the development of world-wide competition for INTEL-SAT contracts.[73]

Having briefly analyzed the major issues of the negotiations leading to the Definitive Arrangements, let us now assess the negotiations from the point of view of American foreign policy. The United States entered the negotiations with a commitment to continue to provide technology which it had largely innovated and developed for the benefit of a world organization composed of 83 states, as of mid-1972. The United States was prepared to relinquish some of the power it had had under the Interim Arrangements, but the results of the negotiations support the conclusion that America made even greater concessions than it had originally planned for. As argued previously, these concessions were not real losses, because in spreading the benefits of communications satellite technology, everyone can gain. As President Richard M. Nixon said in remarks at the end of the conference on May 21, 1971, "there will always be competition between nations, and that competition, if it is peaceful, can be constructive rather than destructive."[74] The President concluded his remarks with these thoughts:

There are many differences in the world which exist today, not because of basic vital interests which are irreconcilable, but simply because of lack of information, because of ignorance, because the people or the governments of one part of the world do not really know the people or the governments of another part of the world.

So as you can well see, this kind of breakthrough, through which it will be possible to have instant communication around the world, will reduce those areas of difference which exist because of ignorance and lack of information to a minimum.

### Distribution and Direct Broadcast Satellites

Let us now inquire into the potential uses, problems, and opportunities which could accompany the introduction of distribution and direct broadcast satellites, an issue not dealt with thoroughly during the negotiations on Definitive Arrangements. We have seen that the problems concerning the entry of direct broadcast into the communications system are different from those surrounding the introduction of point-to-point satellites. Direct broadcast satellites will beam information or programs directly to home radio and television antennas. Control over access will thus depend entirely on the user, assuming he has the equipment and his government does not seek to censor or jam the message. Distribution satellites lie midway between point-to-point and broadcast satellites as they will have the power to send signals to very small receivers, yet not so small as to be within the average home.

Both distribution satellites and direct broadcast satellites offer vistas for mass communications never dreamed of before. Naturally the introduction of these subspecies of communications satellite technology involves different sorts of problems from those relating to point-to-point space communications technology. There is a question not only of disrupting established economic interests but of gauging and handling the political and social effects of the new tools. For instance EUROSPACE, a consortium of European aerospace firms, views the potential ramifications of direct broadcast satellites as more revolutionary than the introduction of nuclear weapons as the following quote indicates:

The possession of a nuclear deterrent by powerful nations dictated world policy during the last two decades, but it has not brought about profound changes in the relationships previously formed in the world. The daily use of the persuasive force of world television broadcasting might in certain circumstances prove to be a dangerously effective political instrument.[75]

What are some of the positive uses that are foreseen for direct broadcast and distribution satellites? One of the most important uses is seen as providing educational services to the developing nations as well as to interested developed states. Given the low literacy rates and the lack of teachers, educators in the less developed countries are severely tasked in their ability to educate their citizens. Education via satellite is seen as a means of leapfrogging the need to train teachers. Television programs on agricultural techniques or birth control devices may even circumvent the lack of literacy. The idea of mass education through direct broadcast satellites received early backing from Hughes Aircraft, the Committee on Communications of the White House Conference on International Cooperation, the Rostow Task Force, etc.[76] In 1969, NASA and India negotiated an agreement to provide instructional television on an experimental basis to 5,000 widely distributed village receiving stations at a cost of only $2.5 to $2.9 million.[77] Unfortunately, this experiment, due to go into operation in 1972, has been delayed till 1974 because of American budget cuts and problems in India. However, indication of the widespread support for educational use of satellites can be found in the fact that the Agency for International Development has been made a focal point for assisting all less-developed countries in communications projects for educational, health, and other purposes.[78] Brazil has been especially interested in a program similar to that with India.[79]

In addition to using distribution and direct broadcast satellites for educational purposes, there are many other possible uses including broadcasting of weather maps for international forecasts, dispatching medical data, linking of the world's libraries, linking (and perhaps influencing the standardization of) the world's computers, and international broadcast of news, sports, and entertainment programs. A list such as this gives an indication of the developments on the horizon, if states can agree to put the new technology to use.

As the estimated technological availability of distribution satellites is at least one year away and of direct broadcast satellites about eight to ten years away, decision makers in various states have considerable lead time to prepare for the introduction of the various services. Technology in satellite communications has not outpaced man's ability to use it for civilized purposes. This conclusion should be contrasted with that of Jean D'Arcy, Director of the Radio and Visual Services Division of the UN Office of Public Information, who states in regards to direct broadcast satellites that "rarely . . . have the scientists and technicians remained so far ahead of the legislators."[80] This observation neglects the fact that within a period of seven months from the appearance of D'Arcy's article, the Definitive Arrangements were initialed, and they do allow for INTELSAT to take on direct broadcast services; it also ignores the fact of the NASA-India agreement.

Let us list briefly some of the major problems that will accompany the technical availability of distribution and direct broadcast satellites. These problems are technical, economic, and political. First of all, there is the major technical and economic problem of frequency allocation. Without an ITU allocation there can be no use of the new satellites, for the frequency spectrum is an extremely scarce resource. A World Administrative Radio Conference of the ITU was held between June 7 and July 17, 1971 to deal with the problems of assigning more frequency space for direct broadcast services, which were previously justified only on an experimental basis. The conference did extend the portion of the frequency spectrum usable for space applications as well as approving direct broadcast service.[81] There still remain other hurdles to global service; for instance, standardization of hardware. A global broadcast capability may not occur due to the controversy over color television.[82] The computers of the world may not be able to carry on conversations with each other unless commercial rivalries are softened. In other words, economic stakes may make the realization of a truly global system impracticable.

Another problem concerns that of censorship or control of the content of international news, educational, and sports events broadcast to mass international audiences. This problem relates to the basic ambivalence concerning international broadcasts—they may be used as instruments of propaganda or as instruments of international understanding. Experience to date with Early Bird, EUROVISION, and INTRAVISION suggests that there has not been a continuing interest in international television programs. There is no dynamically growing demand for this service as there is for overseas channels for telephone.[83] International enthusiasm has built up around certain isolated events such as the

Olympics, the moon landing, or a statesman's funeral, but not about the day to day events peculiar to each culture. Further militating against a widespread use of international broadcasts are differences of language, time, and cultural mores. Another factor is the cost of programming—paid in the United States by advertisers who are familiar with the behavior patterns of their audiences. Who would pay for an international news show to be broadcast to all countries?

In opposition to this pessimistic prognosis, there are those who say that an international broadcast capability will tend to break down language barriers or that language barriers may be circumvented through instantaneous electronic translation. This is clearly a speculative area where existing knowledge gives us no precise information.

The obstacles to global mass communications via satellite have also included the prospect of a continuation of the Cold War. The political (not economic) stakes of establishing service to and within developing nations could cause an intensification of the Cold War rivalry between the United States and the Soviet Union. Such a rivalry might rebound to the benefit of the developing nations if they are able to obtain increased communications capabilities without sacrificing control over their management and content. On the other hand, if either of the great powers were able to sell its system to a particular developing nation, this nation's communications system might then come under the economic and political control of a foreign nation. Foreseeing this danger, it is natural that many of the less developed countries have called for greater international control over direct broadcast systems.[84]

It seems obvious that there must be some form of international control over the content of services beamed from direct broadcast satellites. It takes at least two to communicate, and no state will long permit its citizens to receive potentially disruptive messages from a foreign power. In light of this condition, it will be necessary to reach an international agreement on the uses of direct broadcast satellites before they are orbited. What services from direct broadcast or distribution satellites will be acceptable? How much will they cost? What audiences will be covered? Answers to questions concerning cost and audience will depend on detailed analyses of traffic data, communications networks, and other factors. These studies are now being undertaken by Congress, certain Government agencies, private corporations, other governments, and the United Nations.[85]

Answers to the problem of service selection are more speculative and will depend on political variables. The Committee on Communications of the 1965 White House Conference on International Cooperation proposed that the United Nations establish an agency, the Voice of Peace, to provide informational and educational services to the less developed areas of the world. Satellite communications and other means of communications would be used.[86] However, President Johnson did not and President Nixon has not followed up this suggestion. Rather the United States has chosen to pursue policy via traditional bilateral agreements, as with the NASA-India experiment, while considering INTELSAT's multilateral role in this area as a possible option. The complexities and the economic and political stakes associated with the introduction of direct broadcast satellites would seem to justify this experimental and incremental approach. Indeed, the UN's Working Group on Direct Broadcast Satellites has itself backed a cautious approach.[87] It is imperative upon Comsat, the Government, other governments, and international organizations such as INTEL-SAT, the UN, ITU, and UNESCO to pursue serious studies of the problems and opportunities in this area. There will undoubtedly be clear opportunities for business statesmanship in Comsat and for analyses of the ambivalent aspects of American policy within the Government. Will the laissez-faire approach to commercial programming create an international television wasteland? Would this promote international understanding? Can one promote international understanding without some form of censorship over the uses of the mass media? What will be the concrete program for assuring that American leadership in space communications promotes international peace and understanding?

## Conclusions

The operating experience with INTELSAT and the successful negotiation of Definitive Arrangements plus the technical, political, and economic problems on the horizon concerning direct broadcast satellites gives one an indication of the complexity of the issues confronting United States policy for space communications. It may seem that we are on the threshold of potentially revolutionary developments in satellite communications. But technological developments to date do not appear to merit the label "revolutionary," for their principal disruptive effects have been on the economic investments of certain interests. However, in light of the material from this chapter, it is fitting to ask how technological change could produce revolutionary social effects.

Referring back to the observations of Harold Lasswell, one finds the open-ended proposition that political structures and power relations may be changed by innovations in science and technology but may also dominate these changes.[88] One job of the researcher should be to catalogue different innovations to see where the balance lies. There are three kinds of communications technologies associated with space communications developments: (1) point-to-point, (2) distribution, and (3) direct broadcast. Experience to date has been with point-to-point services. However disruptive this experience may have been to AT&T, the British, and others, the effect on America's general foreign policy has not been profound. It has been incremental. The organizational innovations witnessed by Comsat and INTELSAT have not had significant political consequences on changing nations' expectations concerning international peace and understanding.

Distribution and direct broadcast satellites may promise potentially revolutionary consequences for existing power relations and political structures. The consequences may either come through conscious effort or through unintended consequences of the new technology. Educational services for the less developed countries via satellite could provide these countries with an economic ability to educate their citizens and thus contribute to a narrowing of the gap between the have and the have-not nations. If distribution and direct broadcast services could assist in narrowing the gap, this would be an effect of profound significance. In order to achieve this end, a conscious program would need to be inaugurated. The result would not occur by the unseen hand of technological innovation in the means of communications but through conscious manipulation of the new technologies.

It is possible that one long-range effect of the introduction of distribution and direct broadcast satellites would be to blur the meaning of national boundaries and thus contribute to international peace and understanding. This development would undoubtedly be gradual and subject to reversals by the onset of political conflicts. The very barriers to peace and understanding, which American policy intends satellite communications to overcome, may inhibit the utilization of this new tool. But it is certainly possible that a postindustrial world public opinion might evolve if similar services could be available to each individual in the world.[89] This would indeed be a revolutionary effect and would support the hypothesis that technological change has tended to progressively widen the boundaries of political communities.

In summary, we may say that there can be two revolutionary social effects of the introduction of distribution and direct broadcast satellites: (1) the modernization of the communications networks and educational capacities of the developing nations and (2) the gradual creation of a world public opinion. These effects will not be inevitable results of the new technology but may be social inventions associated with it.

# 9 Conclusions

## Policy Goals and Policy-Making Processes

American policy for space communications is composed of a multiplicity of goals, some of which have received operational clarification and others which reflect ambiguous motives and paradoxical systemic factors. The policy-making processes themselves have not exhibited a uniform pattern. Some issues have been resolved within a stable decision-making system or subsystem and others have involved novel interactions between different systems. Still others have involved the creation of new decision-making arenas.

The question to ask is what role technological change in satellite communications has played in lending precision or ambiguity to the processes and policy outcomes of the processes. The technological change factor, however, is only one variable and it cannot be isolated from other variables responsible for present conditions. Therefore, in order to assess the relative importance of technological change, we will consider all possible causes and then speculate on their relative importance. First, however, let us make a brief survey of the different aspects of policy and the different policy-making patterns which have produced these policies.

It is relatively clear that the United States has: (1) followed a course to promote greater and more efficient use of the frequency spectrum for space communications; (2) promoted the commercial use of space communications for satisfying the increasing demand for channels, first in the North Atlantic area and then in the Pacific and Indian Ocean basins; (3) aided the less developed areas of the world in establishing space communications facilities, especially earth stations; (4) exploited space for military communications; (5) demonstrated its scientific and technological leadership in space communications; and (6) encouraged partnership in INTELSAT by all states including Communist states which are members of the ITU. These accomplishments speak well for the success of American policy for satellite communications as part of the United States' overall foreign policy. Thus, it is clear that the criticisms in 1962 of the foreign policy consequences of the Communications Satellite Act have proved ill-founded in historical retrospect. (See Chapter 4.)

There are some facets of space communications policy which have not benefited from operational clarification. These are: (1) the scope of the goal of leadership; (2) the concrete meaning of the goal of international peace and understanding; and (3) the operational definition of the goal of establishing a

171

single global communications satellite system. While the United States has demonstrated leadership in utilizing space communications for projecting an image of scientific and technological competence and for expediting commercial communications, it has not clarified the uses the new technology will have as diplomatic or psychological instruments of foreign policy. How will leadership in the commercial uses of space communications be related to using space communications as a means of changing those attitudes of man which contribute to war and international tensions? Since mass communications have significant effects on public opinion, and since space communications is a potential instrument for sending programs immediately to global audiences, it is important to ask who will control access to the space communications facilities. Will every individual or state have nondiscriminatory access to send discriminatory propaganda? In short, will American leadership in space communications be limited to providing hardware and commercial service or include initiatives to control harmful effects of mass communications? This is the challenge for the late 1970s. Presently, INTELSAT provides only point-to-point communications rather than a direct broadcast service.

A second ambiguity of American policy is the imprecision surrounding the goal of promoting international peace and understanding. When might the pursuit of this goal produce internally inconsistent action? Is this goal compatible with INTELSAT in which the United States is the senior partner, or would a greater equality of partners promote more fruitful cooperation? If inequality is the reality, how does this condition relate to the achievement of the goals of American foreign policy? If the political advantages of technological change are unequally distributed among states, how can this situation be overcome in order to promote cooperation among equals?

These questions are related to the ambiguity connected with the goal of establishing a single global system. The selection of this goal implies that the alternative of several regional and international systems serving the whole globe is unacceptable. Why? It seems apparent that a global system serving commercial, cultural, political, educational, and other purposes will have to have different sets of rules relating to different sorts of traffic. There will also be different technical capabilities associated with different services, each requiring separate frequency space.

What, then, does the single global-system goal imply? It could mean that there would be one system of ownership and management for all services. It could mean that the same set of rules should apply to any one particular service, although there would be different rules for disparate services. If one global system will "avoid destructive competition to tie different countries into the communications systems of different blocs,"[a] it would seem to be a system including Communist countries. If the Communist countries establish INTER-SPUTNIK and other regions set up economically competitive but comple-

---

[a]See p. 85.

mentary operations, what kind of a failure would this be for American foreign policy? If the United States wants to avoid destructive competition in space communications, why not in international cables and radio too? Is the single global-system idea an unhappy parallel with over-emphasizing the unity of NATO and the "Free World"? Perhaps the United States has learned a lesson from the days when the world was seen in tones of black and white, for America accepted the Definitive Arrangements, which, while committing members to a single system, allows for separate international systems (Article XIV(d)(e)).

There are, then, ambiguities in the substance of American foreign policy for satellite communications. In addition, one may notice that the policy-making processes have often been lacking in central direction and clear delineation of jurisdictional authority. One may see this in four decision processes relating to space communications issues:

The first case involves the ownership issue on which, in 1961, the FCC and NASA roles were criticized by the Antitrust Division of the Department of Justice. In Chapter 3 we noted that the basic issue in question was whether the established carriers would be given ownership of satellite communications because of their experience or whether a new arrangement would be worked out more in keeping with the antitrust laws. It was shown how the President and the Congress gradually became involved in the policy-making process, although there was no certainty beforehand that this would be the case. With the expansion of the policy-making arena, there was a proliferation of issues in addition to the antitrust issue. We noted that policy-making was legislative in the sense that the participants were relatively equal in power; there were disagreements about goals; and there existed numerous alternatives.[1] What we examined was a problem area where issues transcended and transformed the traditional communications environment, which had been characterized by low-visibility interaction between the FCC and the international carriers.

A second example of legislative politics can be seen in the frequency allocation issue on which, in 1963, there was international problem-solving. In contrast to the first example, American policy for frequency allocations was made in close harmony with other nations. In other words, legislative policy-making was characteristic of international politics and American foreign politics, not just domestic politics. Although there was legislative politicking, the intensity of the conflict was less than in the first case. There was mutual adjustment of views both within the United States and between the United States and other countries. In fact, it was noted that the conflicts in allocating spectrum space were often not between countries but between functional groups of users. Accommodation of states and users took place according to an established pattern of decision-making within a stable international subculture.

A third example of lack of central direction in policy-making can be seen in the case of the international organization issue on which, in 1963, there was much early confusion of views and the roles of the Government, Comsat, and potential European partners. The initial lack of clarity threatened to result in severe crises. There was more misunderstanding between Comsat and the Government than between both these participants and the governments of foreign countries; the State Department's views were closer to those of the Europeans' than to those of Comsat's leadership. In fact, it was the role of the ECSC to bring harmony to American policy-makers. Consultations with foreign governments were an important factor contributing to the substance of United States foreign policy. A joint Comsat-State team then followed this policy in negotiating with these foreign governments to establish INTELSAT. It is interesting to note that the resolution of the ownership issue in the international negotiations was an easier task than the resolution of the ownership issue for the establishment of a domestic communications satellite system is proving to be.

A fourth example of policy-making through the mutual adjustment of participants of equal authority can be seen in the case of the Defense Department's negotiations with Comsat for a military communications satellite system. Since this involved urgent national security requirements, one might expect that policy would be made secretly, quickly, and hierarchically. This is the way the process started, but it ended openly, slowly, and legislatively. The House Military Operations Subcommittee took exception to the Defense Department's arrangements with Comsat, questioning whether the Comsat contract would in fact fully satisfy urgent needs, whether there might not be a question of subsidy to a Government-sponsored corporation, and whether Defense negotiations with Comsat might not be inconsistent with Comsat's role in INTELSAT. The result of the extensive and intensive investigation by the Subcommittee was to scuttle a Defense contract with Comsat. Gradually, as technical, economic, and political issues were clarified, Defense returned to its initial contract with Philco.

We have pointed to six policy goals which are relatively clear and three which are ambiguous. We have briefly mentioned four examples of policy-making processes which were legislative rather than executive. Is there any relation between the processes and the outcome? It appears that clear and precise goals have come from a "muddling through" process, whereas ambiguous goals have resulted from what one might think should be a highly centralized process. The goals of leadership, peace and understanding, and a single global communications satellite system have received little operational precision. Why is this? Answering these questions involves a consideration of the interaction of three processes— the three themes of the book.

## "Muddling Through"

The policy process is often characterized by minimal central direction, yet policy outcomes themselves do not necessarily suffer, judged by the statements of participants. Why is this? It is because, as Charles E. Lindblom has argued, incremental decision-making through incremental adjustment of partisan decision-makers gives scope for accommodating competing values and promoting shared values. It is impossible to achieve a completely clear view at any one time. Policy-makers are often engaged in sequential learning and revising, and it is best to recognize this by not overcentralizing the management of policy in the hopes that long-range planning will be a panacea for the disadvantages of pragmatic decision-making. The cure may be worse than the disease. Or, as Kenneth Boulding has written:

Decision-making by instinct, gossip, visceral feeling, and political savvy may stand pretty low on the scale of rationality, but it may have the virtue of being able to take in very large systems in a crude and vague way, whereas the rationalized processes can only take subsystems in their more exact fashion, and being rational about subsystems may be worse than being not very rational about the system as a whole.[2]

Failure to clarify the meaning of the goals of leadership, international peace and understanding, and the establishment of a single global system may also result from the psychological ambivalence of the policy-makers themselves. We have seen that ideas on national security are not altogether compatible with promoting an equal partnership, even with America's own allies. The goal of one global system may express desires for leadership as well as for promoting fruitful cooperation. In these instances, imprecise policy is a cover for a schizophrenic attitude or mood reaction on the part of the policy-makers as a result of the fact that they live in an international system where a purely cooperative policy might undermine national security and a deterrent policy involves paradoxes.[3]

One may say that the ambiguities in American satellite communications policy are results of three factors: (1) a reasonable approach to policy-making in a complex, pluralistic, and democratic society where possibilities for action occur at different intervals; (2) certain deficiencies of characteristics in the psychological makeups of the important decision-makers; and (3) basic security dilemmas arising from the dynamics of the international political system. It is probable that all three factors play a part in explaining the lack of clarity in certain goals of American foreign policy. Since we are not making a laboratory experiment, we must speculate on the relative importance of the three factors.

The leadership goal may be construed as a case of ambivalent personality traits involving a desire to excel for reasons of pride and prestige combined with a desire to have political power over others. Yet the desire to have power over other states may conflict with the goal of sharing technological capabilities in order to attain common values. Hence, the goal of contributing to international peace and understanding may suffer; but, if the United States were to promote cooperation in a purely altruistic sense, it might in turn have to contend with another state's or region's desire to gain political predominance with the assistance of space communications techniques.

The goal of establishing a single global system also must await the development of policy by other nations and INTELSAT, as well as the United States, before it can be seen in sharper focus. It does seem, however, that American negotiators for the Definitive Arrangements gave greater attention to the ambiguities of this goal during 1969-1971 than was the case during the 1961-1962 debate. It became clear that the one-global system goal was really a means for achieving other goals. Seen in this light, this subgoal was reevaluated in the context of weighing the advantages and disadvantages of having several economically competitive systems serving complementary purposes. The Definitive Arrangements, while still ambivalent in this regard, do provide options for complementary international, regional, and domestic systems.

In summary, one may say that American policy for satellite communications has proved a success by its own standards. There are failings in policy but these relate to the achievement of ends which are inherently paradoxical or must be achieved in future years. Although the policy process has been fragmented both within the Government and Congress, the dispersal of responsibilities and the overlapping of jurisdictions have not proved to be insuperable obstacles to the formation, successful operation, and negotiation of Definitive Arrangements for INTELSAT. In fact, the great multiplicity of decision-makers has been a boon for American policy in an area where problem-solving cannot be synoptically accomplished in one panoramic vision but must be strategically pursued over time.[4]

### The Impact of Technological Innovation

Related to the fact that the complex matrix of space communications policies is being formulated and implemented in a pragmatic and incremental fashion, is the fact that technological innovation in space communications has blurred the boundaries of existing communications arenas. The introduction of the new technology in space communications has made possible the creation of new participants, Comsat and INTELSAT, and has modified the activities and responsibilities of old participants such as the FCC and AT&T. The fact that the participants in the making and resolving of various problems connected with the introduction of the new technology into the traditional environment were unclear as to how the new technology would affect them—or how they wanted to control the technology—added complexity to the policy-making process. The delivery of one, concise, clear policy statement from the Government was thus unlikely. One impact of technological change in satellite communications is to lend further support to the arguments for the rationality of incremental decision-making. Improvisation from year to year in an environment which is changing is more rational than imposing a long-range solution which would probably become unworkable in light of future developments.

This point is a contribution to Professor Lindblom's analysis. He does not make the proposition, which this study bears out, that incremental decision-making is more appropriate for environments characterized by continuous technological innovation. The effect of technological innovations in communications satellites was not to change the nature of incremental decision-making but to be another factor making for complexity and thus justifying "muddling through."

Perhaps, however, this conclusion relates only to technological changes which do not have revolutionary impacts. One may want to promote long-range planning and central authority for dealing with technological changes which may have certain unforeseen side effects which could be extremely harmful or beneficial—such as cheap birth control pills, supersonic transports, or antimissile missiles.

One may also want or see central management in environments where, as Samuel P. Huntington writes, "1) the participating units differ in power (i.e., are hierarchically arranged); 2) fundamental goals and values are not at issue; and 3) the range of possible choice is limited."[5] Such environments are stable for the reasons cited. Varying these factors would produce more or less stability.

For instance, if fundamental goals were at issue and there was a wide range of choice, it is still possible to imagine a highly centralized power structure maintaining an executive policy-making process. Such might be the case with the choice of when and how leaders might introduce mass communications into developing nations. Thus, technological innovation might have effects on policy and society but not on the policy-making process.

Stable technological environments may have either stable legislative or executive policy-making patterns depending on the characteristics of the variables Huntington cites. The stability of the technology would appear to be only one variable explaining the stability of the process—and not the most important one at that.

If technology is stable and the problem is purely technical, then decision-making can be made by machines as with logistics and certain inventory problems.[6] There is no politics. But where there is politics, even if technology is stable, the type of process cannot be predicted. It is thus illegitimate to draw general conclusions about the effect of technological innovation on innovations in policy-making machinery and processes.

To date, innovations in satellite communications technology could not be said to compel changes in the expectations or behavior of men. The introduction of satellite communications technology to the world did not determine the political and economic attitudes of governments, carriers, and manufacturers. But as a social invention, it did influence their attitudes and was in turn influenced by them.

The availability of operational communications satellites was widely viewed as a long-range prospect in 1957 and 1958 but as a short-range or immediate prospect from 1961 through the Comsat debate of 1962. Much of the reason for the switch in expectations was a desire by certain carriers and manufacturers to enter the field before their competitors, so as to establish a commanding position. Thus, the motive of economic profit played a large part in changing expectations, and this in turn affected the flow of research and development funds. The change was also due to the political motives of the Kennedy Administration which wished to demonstrate leadership in space as against the Soviet Union.

The feeling of urgency subsided after major economic questions had been decided by the creation of Comsat as a potential competitor to the traditional carriers, and after it became apparent that the Soviets were behind the United States in orbiting communications satellites. Whereas a low-orbit system could have been launched by 1963 or 1964, it was decided that a truly operational, efficient system could not be available until 1967 or 1968. This expectation changed in late 1965 and early 1966 when NASA decided it wanted Comsat and INTELSAT to provide communications for the Apollo Program. The excess capacity in this system could have made a nearly global capability available by late 1966. However, problems with the Apollo Program in 1966 and 1967 changed this expectation and INTELSAT did not deliver global service until the summer of 1969.

In this shifting pattern of expectations, one sees not the inevitable force of technology making its mark on man but a multiplicity of economic, political, and technological influences operating on each other. This observation is related to the evolutionary or revolutionary characteristics of technological change. We have seen that the term "revolutionary" was often used by certain corporations to support their economic interests against established interests in international communications such as AT&T and ITT. These latter interests considered space communications to be an engineering application—a microwave relay station in the sky.[b]

It has been emphasized that the disruptive effects of the introduction of satellite communications have been limited so far to the economic sphere. The interests associated with cables and microwave relay towers have suffered because of the creation of Comsat as a competitor. Even here, however, the economic consequences of satellite communications have not been destabilizing because operations are still in their early stages. It is clear that communications facilities will be overhauled with the growing demand for channels and the ability of communications satellites to fill the gap. But there is no formula to the exact mix of the old with the new, or to the economies which may be associated with developments in lasers and waveguide cables. It is clear, however, that the characteristics of the new environment will be influenced by the attempts of established interests to control new technologies within the traditional framework, and the attempts of other interests to promote new values. What one sees

[b]See Chapter 3.

is the old FCC-international carrier subsystem with Comsat as a new member of the club.

On the other hand, in the late 1970s we may see some surprising innovations in the structure and functions of international communications. It is possible that developments in satellite communications will have politically revolutionary consequences. If the new technology is used as part of a conscious program to upgrade the educational systems of the developing nations, one effect of space communications could be to narrow the gap between the developed and less developed countries. Handled in the proper fashion, space communications techniques could lead to greater awareness on the part of citizens of traditional societies. The creation of informed public opinion and mass democracy have depended upon advances in the means of communications—the press, radio, movies, and television.

Satellite communications could also contribute to greater individuality both in the developing and so-called advanced societies. As part of a future visionary system where each individual family has its own computer, satellite communications could provide the needed services and frequencies to allow every home access to information and programs of its own choosing. Technology would then be associated with decentralizing tendencies in a postindustrial society rather than centralizing forces.

There may also be long-run, unintended revolutionary effects of space communications, for instance, in influencing the development of world public opinion. But this is a very tenuous possibility, given existing knowledge on the effects of mass communications and the probability that present differences will dominate the capability of a space communications system to beam similar programs to world audiences. If a world public opinion were to evolve in a gradual manner, this would be an instance of technological change operating as a "force," that is, having unintended side-effects.

There have been many novel features associated with developments in space communications, but this observer does not see them as revolutionary. Comsat and INTELSAT have been established to cope with the new problems and opportunities, but they are also adaptive mechanisms to assure that the new technology does not overly destabilize existing arenas of power. This observation is supported by the operating experience of Comsat and INTELSAT.

In the United States there is a widespread tendency to see technological change as revolutionary and also beneficial. This outlook is probably the result of the fact that so many novel developments in technology have been beneficial as well as having unforeseen and pleasing side-effects. The tendency to see technology in an optimistic glow has been countered in recent decades by observers who see some areas of the new technology, for instance automation and cybernation, as alienating man. These observers see gaps or lags between the availability of the new techniques and man's ability to control them for civilized purposes. Technological change has acted as a force beyond man's control, whereas it should be controlled to promote man's values.

By reifying technological change, optimists and pessimists miss the point. The effects of technological change have been ambivalent. As features of a heterogeneous society with many competing and overlapping values, the characteristics of technological change are more understandable. The TV wasteland of some is the happy recreation of others. The automated alienation of some is opportunity for leisure to others. The inevitable nuclear conflict of certain observers is the system of stable deterrence to others.

Space communications innovations are beneficial to certain interests and harmful to others. The timing and manner of the introduction of types of communications satellite technology have been significantly influenced by the political and economic process of mutual adjustment, characteristic of the democratic process within the United States, and of commercial-diplomatic negotiations. There have been innovations in communications policy as a result of combining space policy with traditional communications policy, and the underlying condition for this development has been technological innovation. However, the innovations are not revolutionary. Revolutionary innovations have often been the result of crisis situations, and, in this case, there have not been any real crises, although some have been contrived. This observer has been impressed by the fact that, up to the early 1970s, space communications technology has been subordinated to existing political processes and structures rather than modifying them. In addition, the new technology has to a large degree been used as a servant of the goals of American policy rather than deforming these ends.

### The Boundaries of Domestic and Foreign Policy

Related to the complexity of the interests affecting the development of space communications technology and to the pluralistic character of the policy-making arena, is the problem of drawing a distinction between the substance of foreign and domestic policy, and also between the procedures of foreign and domestic policy-making. American space communications policy is a matrix of policies, some of which are addressed principally to the solution of problems affecting domestic interests and others of which are directed primarily to achieving certain goals in cooperation with foreign nations. Yet there is a wide frontier where the consequences of resolving certain issues affects both environments.[c] Consequently, the process of making decisions on these types of issues sometimes involves participants from both environments.

One example of an issue which breaks down the distinction was the ownership issue—the major problem to be resolved during the 1962 Congression-

[c]See pp. 44-45. This breakdown should be reformulated for succeeding years. For instance, since 1961 the subsidy issue has moved into the second category. So has the problem of frequency allocations for broadcast satellites.

al debate. The creation of Comsat added an additional participant to the list of international carriers and potentially to domestic carriers, as the economic position of space communications in the future domestic communications network makes Comsat's stake in this field even more important than in the international network. Yet Comsat was established as an instrument of foreign policy to achieve certain far-reaching national goals. Given the Corporation's role as a profit-making company, as an instrument of foreign policy, and as a member and manager of INTELSAT, it is understandable why the establishment of Comsat was a decision of both domestic and foreign policy.

As Congress' role in foreign policy-making is often played down,[7] it is important to keep counter-evidence in view so as not to offer partial explanations. The evidence from the 1962 Comsat decision leads to two conclusions. Congress played a greater role here than it has in situations of crisis such as Berlin or the Cuban missile crisis. Noncrisis situations thus afford Congress or its committees more opportunity to play a substantial role in policy-making.[8] (It may also be postulated that Congress' role is greater in chronic crisis situations, e.g., pollution, than in immediate crisis situations such as those referred to above.)

Secondly, since the establishment of Comsat involved issues of both domestic and foreign policy, it is inappropriate to draw a sharp distinction between foreign and domestic policy-making in the first place. This is true of other decisions such as the SST, tariffs, and foreign aid. These "national" decisions should be the subject of studies directed to comparing and contrasting the politics of different issue areas.[9]

These same arguments apply to interest group influence in foreign policy-making. Manufacturing and carrier interests played and do play more of a role in making foreign policy for satellite communications because it is a non-crisis issue area. Secondly, many of these corporations are multinational in character and, thus, to varying extents, transnational interest groups. Therefore, it is meaningless to say that interest groups play less important roles in determining foreign than domestic policies if what we are really viewing is the politics of a penetrated system.

Another example of where domestic politics has overlapped foreign politics is seen in the supervisory role of various Government agencies, the FCC being the prime example. The Commission is usually pictured as a regulator of domestic common carriers with only a minor role to play in regulating international traffic via cable and radio. Space communications is in a different category, however. The Commission's novel role in regulating the procurement practices of Comsat involves the Commission in the foreign policy process, as Comsat is a representative of an international organization—INTELSAT—besides being an American corporation. In 1966, the FCC impinged on INTELSAT's domain in the guise of regulating Comsat.[d] The FCC is also part of the foreign policy process in

---

[d]See pp. 142-43.

deciding on issues such as earth station ownership and cooperation with Canada in establishing her domestic communications satellite system.

While it is not particularly exceptional that resolutions of certain types of issues affect both foreign and domestic policy, it is novel that the process of making decisions on some of these issues has involved both domestic participants and foreign governments. This was the case in the making of American foreign policy for negotiating the Interim and Definitive Arrangements. The process of reaching a cohesive position within the United States was promoted by discussions with foreign governments. The policy-making process of INTEL-SAT's ICSC has also reflected the interpenetration of the domestic and foreign policy environments. For instance, decisions on rates for space segment use by the ICSC necessarily affect AT&T's and ITT's stake in cables. Rather than appealing to the FCC as the highest domestic arbiter, the established international carriers either have to accept INTELSAT's decision, or work within Comsat to influence its stand on the ICSC or Board of Governors. The decision-making process thus takes place in several different environments. This is what James N. Rosenan calls the "penetrated" system.[e]

The process of decision-making on problems of satellite communications has tended to be legislative rather than hierarchical-executive. This is true of foreign policy, national security policy, international decision-making as witnessed by the EARC of 1963,[f] the Defense Department-Comsat discussions of 1963-1964,[g] the establishment of the Interim and Definitive Arrangements,[h] and the operations within INTELSAT.[i] Policy-making has been characterized by widespread participation and discussion of the issues by diverse interests within and outside the United States.

The explanation of the legislative character of the decision-making process may be found in two factors. Within the United States responsibilities are dispersed and the intricacies of making viable decisions necessitate a process of mutual accommodation. This is also true of decisions on international matters such as frequency allocations, and this is related to the nature of the communications arena. It takes at least two to communicate, so, understandably enough, there has to be mutual coordination. This situation is not in general true of decision-making on matters involving the technologies of production and military coercion, although it tends to be true of transportation.

The second factor explaining the legislative character of policy-making in satellite communications is that the participants frequently communicate with each other, and this contributes to a mutuality of expectations and greater understanding between actors. We saw in Chapter 5 that Comsat's officials have

[e]See p. 44.

[f]See Chapter 5.

[g]See Chapter 6.

[h]See Chapters 5 and 8.

[i]See Chapter 8.

had varied backgrounds with the Government, the carriers, and other corporations. This has contributed to operational understanding of the complexities and nuances of the environment. There is no indication of dogmatic ideas on free enterprise or socialism. This mutual clarity of expectations at the domestic level is true of the working of INTELSAT on the international level. This is natural, given the monthly meetings of the ICSC and the Board of Governors and the frequent interaction on the subcommittees. International decision-making is taking place within the framework of an ongoing international organization rather than in isolated diplomatic negotiations.

The intertwining of many domestic and foreign policies and policy-making processes is related to the scope and complexity of the problems and opportunities associated with the innovation of communications satellites. One can expect this interaction of environments to continue as man seeks to control the introduction of distribution and direct broadcast satellites. One can expect that the clarity of American policy will be more apparent not after the United States simplifies and centralizes its organization for space communications (and perhaps all communications), but through recognition of the complexities of the changing environment and the necessity for transitory solutions to emerging problems and opportunities.

# Appendixes

## U.S. and INTELSAT Communications Satellites 1958-1972

| Date | Name | Launch Vehicle | Remarks |
|---|---|---|---|
| Dec. 18, 1958 | Score | Atlas B | First Comsat, carried taped messages. |
| Aug. 12, 1960 | Echo I | Thor-Delta | 100-foot balloon served as first passive Comsat, relayed voice and TV signals. |
| Oct. 4, 1960 | Courier 1B | Thor-Able Star | First active-repeater Comsat. |
| Mar. 30, 1961 | Lofti I | Thor-Able Star | Low-frequency experiment; failed to separate from rest of payload. |
| Oct. 21, 1961 | Westford I | Atlas-Agena B | First attempt to establish filament belt around earth; failed to disperse as planned. |
| Dec. 12, 1961 | Oscar I | Thor-Agena B | First amateur radio "ham" satellite. |
| June 2, 1962 | Oscar II | Thor-Agena B | |
| July 10, 1962 | Telstar I | Thor-Delta | Industry-furnished spacecraft in near-earth orbit. |
| Dec. 13, 1962 | Relay I | Thor-Delta | Active-repeater Comsat. |
| Feb. 14, 1963 | Syncom I | Thor-Delta | Successfully injected into near-synchronous orbit but communication system failed at orbital injection. |
| May 7, 1963 | Telstar II | Thor-Delta | |
| May 9, 1963 | Westford II | Atlas-Agena B | Filaments formed reflective belt around earth as planned for emergency communications experiment. |
| July 26, 1963 | Syncom II | Thor-Delta | First successful synchronous orbit active-repeater Comsat. After experimental phase, used operationally by DOD. |
| Jan. 21, 1964 | Relay II | Thor-Delta | |
| Jan. 25, 1964 | Echo II | Thor-Agena B | 135-foot balloon, passive Comsat, first joint use by United States and U.S.S.R. |

| Date | Name | Launch Vehicle | Remarks |
|---|---|---|---|
| Aug. 19, 1964 | Syncom III | Thor-Delta | Synchronous-orbit Comsat; after experimental phase, used operationally by DOD. |
| Feb. 11, 1965 | LES I | Titan IIIA | Experimental payload did not reach intended apogee. |
| Mar. 9, 1965 | Oscar III | Thor-Agena D | |
| Apr. 6, 1965 | Intelsat I (Early Bird) | Thor-Delta | First Intelsat (Comsat Corporation) spacecraft, 240 2-way voice circuits; commercial transatlantic communication service initiated June 28, 1965. |
| May 6, 1965 | LES II | Titan IIIA | All solid state advanced experiment. |
| Dec. 21, 1965 | LES III | Titan IIIC | All solid state, UHF signal generator. |
| | LES IV | | All solid state, SHF or X band experiment. |
| | Oscar IV | | |
| June 16, 1966 | IDCSP 1-7 | Titan IIIC | Initial defense communication satellites program (IDCSP)-Active-repeater spacecraft in near-synchronous orbit, random spaced. |
| Oct. 26, 1966 | Intelsat II-F1 | Thor-Delta(TAT) | First in Intelsat II series spacecraft; 240 2-way voice circuits or 1 color TV channel. Orbit achieved not adequate for commercial operation. |
| Nov. 3, 1966 | OV 4-1T OV 4-IR | Titan IIIC | Transmitter and receiver for low-power satellite-to-satellite F layer experiments. |
| Dec. 7, 1966 | ATS I | Atlas-Agena D | Multipurpose, including VHF exchange of signals with aircraft. |
| Jan. 11, 1967 | Intelsat II-F2 | Thor-Delta(TAT) | Transpacific commercial communication service initiated Jan. 11, 1967. |
| Jan. 18, 1967 | IDCSP 8-15 | Titan IIIC | |
| Mar. 22, 1967 | Intelsat II-F3 | Thor-Delta(TAT) | Positioned to carry transatlantic commercial communication traffic. |

| Date | Name | Launch vehicle | Description |
|---|---|---|---|
| Apr. 6, 1967 | ATS II | Atlas-Agena D | Multipurpose, but did not attain planned orbit. |
| July 1, 1967 | IDCSP 16–18 | Titan IIIC | |
| | LES V | | Tactical military communications tests with aircraft, ships, and mobile land stations from near synchronous orbit. |
| | DATS | | Electronically despun antenna experiment. |
| | DODGE | | Multipurpose, gravity stabilized. |
| Sept. 27, 1967 | Intelsat II–F4 | Thor-Delta(TAT) | Positioned to carry commercial transpacific communication traffic. |
| Nov. 5, 1967 | ATS III | Atlas-Agena D | Multipurpose including communications. |
| June 13, 1968 | IDCSP 19–26 | Titan IIIC | |
| Aug. 10, 1968 | ATS IV | Atlas-Centaur | Multipurpose; failed to separate from Centaur, did not reach planned orbit. |
| Sept. 19, 1968 | Intelsat III(F-1) | Thor-Delta(TAT) | Failed to achieve orbit. |
| Sept. 26, 1968 | LES 6 | Titan IIIC | Continued military tactical communications experiments |
| Dec. 18, 1968 | Intelsat III(F–2) | Thor-Delta(TAT) | First in Intelsat III series of spacecraft, 1,200 2-way voice circuits or 4 color TV channels. Positioned over Atlantic to carry traffic between North America, South America, Africa, and Europe. Entered commercial service on Dec. 24, 1968. |
| Feb. 6, 1969 | Intelsat III(F–3) | Thor-Delta(TAT) | Stationed over Pacific to carry commercial traffic between the United States, Far East, and Australia. |
| Feb. 9, 1969 | Tacsat I | Titan IIIC | Demonstrated feasibility of using a spaceborne repeater to satisfy selected communications needs of DOD mobile forces. |
| May 22, 1969 | Intelsat III(F–4) | Thor-Delta(TAT) | Stationed over Pacific to replace F–3 which was moved westward to the Indian Ocean. Completes global coverage. |
| July 26, 1969 | Intelsat III(F–5) | Thor-Delta(TAT) | Spacecraft failed to achieve the proper orbit. Not usable. |
| August 12, 1969 | ATS V | Atlas Centaur | Multipurpose; for millimeter and L band communications; entered flat spin. |

| Date | Name | Launch Vehicle | Remarks |
|---|---|---|---|
| Nov. 22, 1969 | Skynet I (IDCSP–A) | Thor-Delta(TAT) | Launched for the United Kingdom in response to an agreement to augment the IDCSP program. |
| Jan. 15, 1970 | Intelsat III(F–6) | Thor-Delta(TAT) | Stationed over Atlantic to carry commercial traffic between the United States, Europe, Latin America, and the Middle East. |
| Jan. 23, 1970 | Oscar V (Australia) | Thor-Delta(TAT) | Ham radio satellite built by amateur radio operators at Melbourne University, Melbourne, Australia. |
| Mar. 20, 1970 | NATOSAT–I (NATO–A) | Thor-Delta(TAT) | First NATO satellite, stationed over Atlantic to carry military traffic between the United States and other NATO countries. |
| Apr. 23, 1970 | Intelsat III(F–7) | Thor-Delta(TAT) | Stationed over Atlantic to carry commercial traffic between the United States, Europe, North Africa, and the Middle East. |
| July 23, 1970 | Intelsat III(F–8) | Thor-Delta(TAT) | Spacecraft failed to achieve the proper orbit. Nor usable. Last launch of Intelsat III series. |
| Aug. 22, 1970 | Skynet II (IDCSP–B) | Thor-Delta(TAT) | Launched for the United Kingdom in response to an agreement to augment the IDCSP program. Spacecraft failed to achieve the proper orbit. |
| Jan. 26, 1971 | Intelsat IV(F–2) | Atlas-Centaur | First in Intelsat IV series of spacecraft; 3–9,000 2-way voice circuits or 12 color TV channels. Positioned over the Atlantic. |
| Feb. 3, 1971 | NATOSAT–II (NATO–B) | Thor-Delta(TAT) | Second NATO satellite, stationed over the Atlantic to carry military traffic. |
| Nov. 3, 1971 | DSCS 2–1, 2 | Titan IIIC | Operational defense communications satellites launched as a pair to 24-hour synchronous orbits to provide high capacity voice, digital, and secure voice communications for military networks. |
| Dec. 19, 1971 | Intelsat IV(F–3) | Atlas Centaur | Second in new high-capacity series. Atlantic Ocean. |
| Jan. 22, 1972 | Intelsat IV(F–4) | Atlas Centaur | Third in new high-capacity series. Pacific Ocean. |
| June 13, 1972 | Intelsat IV(F–5) | Atlas Centaur | Fourth in new high-capacity series. Indian Ocean. |

Source: Aeronautics and Space Report of the President (1971), pp. 117-18.

**Soviet Communications Satellites**

| Name | Date | Apogee | Perigee (km) | Inclination (degrees) | Period (minutes) |
|------|------|--------|--------------|----------------------|------------------|
| Molniya 1-1 | 4/23/65 | 39,957 | 548 | 65.0 | 720.0 |
| Molniya 1-2 | 10/13/65 | 40,000 | 500 | 65.0 | 719.0 |
| Molniya 1-3 | 4/25/66 | 39,500 | 499 | 64.5 | 710.0 |
| Molniya 1-4 | 10/20/66 | 39,700 | 485 | 64.9 | 713.0 |
| Molniya 1-5 | 5/24/67 | 39,810 | 460 | 64.0 | 715.0 |
| Molniya 1-6 | 10/3/67 | 39,600 | 465 | 65.0 | 712.0 |
| Molniya 1-7 | 10/22/67 | 39,740 | 456 | 64.7 | 714.0 |
| Molniya 1-8 | 4/21/68 | 39,700 | 460 | 65.0 | 713.0 |
| Molniya 1-9 | 7/6/68 | 39,700 | 470 | 65.0 | 715.0 |
| Molniya 1-10 | 10/5/68 | 39,600 | 490 | 65.0 | 712.0 |
| Molniya 1-11 | 4/11/69 | 39,700 | 470 | 65.0 | 713.0 |
| Molniya 1-12 | 7/22/69 | 39,540 | 520 | 64.9 | 711.0 |
| Molniya 1-13 | 2/19/70 | 39,175 | 487 | 65.3 | 703.0 |
| Molniya 1-14 | 6/26/70 | 39,280 | 470 | 65 | 705.0 |
| Molniya 1-15 | 9/29/70 | 39,300 | 480 | 65.5 | 706.0 |
| Molniya 1-16 | 11/27/70 | 39,430 | 435 | 65.3 | 707.0 |
| Molniya 1-17 | 12/25/70 | 39,600 | 480 | 65 | 712.0 |
| Molniya 1-18 | 7/28/71 | 39,300 | 470 | 65.4 | 705.0 |
| Molniya 2-1 | 11/24/71 | 39,350 | 460 | 65.4 | 706.0 |
| Molniya 1-19 | 12/20/71 | 39,200 | 490 | 65.5 | 703.0 |
| Molniya 1-20 | 4/4/72 | 39,260 | 480 | 65.6 | 695.0 |
| Molniya 2-2 | 5/19/72 | 39,300 | 460 | 65.5 | 705.0 |

Source: Science Policy Research Division, Library of Congress.

# Appendix B

## International Telecommunications Satellite Consortium Members, Quotas and Utilization

(at December 31, 1970)

As of July, 1972 there were 83 members of INTELSAT rather than the 77 on this list. The six new countries are Barbados, Costa Rica, Gabon, Ghana, Malagasy Republic, and Mauritania. In each case, the investment quotas of these countries is less than .1 percent.

| Country | Entity | Entry into Effect of Interim Agreement | Investment Quota | Percentage of Total Use |
|---|---|---|---|---|
| Algeria | Ministry of Posts and Telecommunications | February 19, 1965 | 00.539 | – |
| Argentina | Empresa Nacional de Telecommunicaciones (ENTEL) | May 19, 1965 | 01.402 | 1.84 |
| Australia | Overseas Telecommunications Commission | August 24, 1964 | 02.372 | 2.46 |
| Austria | Bundesministerium für Verkehr und Elektrizitätswirtschaft, Generaldirektion für die Postund Telegraphenverwaltung | May 6, 1965 | 00.173 | .18 |
| Belgium | Régie des Télégraphes et des Téléphones | February 10, 1965 | 00.949 | .94 |
| Brazil | National Telecommunications Council | May 17, 1965 | 01.402 | 1.68 |
| Cameroon | Ministry of Transport, Posts and Telecommunications | November 6, 1969 | 00.050 | – |
| Canada | Canadian Overseas Telecommunication Corporation | August 20, 1964 | 03.234 | 3.15 |
| Ceylon | Permanent Secretary in charge of Ministry of Posts and Telecommunications of Ceylon | February 17, 1965 | 00.045 | – |
| Chile | Empresa Nacional de Telecomunicaciones S.A. | May 18, 1965 | 00.280 | 1.04 |

| Country | Entity | Entry into Effect of Interim Agreement | Investment Quota | Percentage of Total Use |
|---------|--------|----------------------------------------|------------------|-------------------------|
| China | Directorate General of Telecommunications of the Republic of China | February 17, 1965 | 00.090 | .94 |
| Colombia | Government of Colombia | February 19, 1965 | 00.539 | .65 |
| Congo, Democratic Republic of | Office Congolais des Postes et Télécommunications | February 2, 1970 | 00.050 | – |
| Denmark | Generaldirektoratet for Post og Telegrafvesenet | March 3, 1965 | 00.345 | .58 |
| Dominican Republic | Compañía Dominicana de Teléfonos, C. por A. | January 12, 1970 | 00.050 | .55 |
| Ecuador | Dirección General de Telecomunicaciones del Ecuador | October 28, 1970 | 00.500 | .14 |
| Ethiopia | Government of Ethiopia | February 19, 1965 | 00.072 | – |
| France | Government of the French Republic | January 18, 1965 | 05.261 | 2.39 |
| Germany | Deutsche Bundespost | September 21, 1964 | 05.261 | 2.97 |
| Greece | Greek Ministry of Communications Directorate General of Telecommunications | May 19, 1965 | 00.093 | .74 |
| Guatemala | Empresa Guatemalteca de Telecomunicaciones | March 7, 1969 | 00.050 | – |
| India | Goverment of India | February 19, 1965 | 00.467 | – |
| Indonesia | Denwan Telekomunikasi | February 19, 1965 | 00.269 | .55 |
| Iran | Ministry of Posts, Telegraph and Telephone | September 3, 1968 | 00.248 | .44 |
| Iraq | Ministry of Communications of Iraq | February 17, 1965 | 00.009 | – |
| Ireland | An Roinn Poist Agus Telegrafa | October 5, 1964 | 00.302 | .28 |
| Israel | Ministry of Posts State of Israel | November 30, 1964 | 00.564 | .41 |

| Country | Entity | Entry into Effect of Interim Agreement | Investment Quota | Percentage of Total Use |
|---|---|---|---|---|
| Italy | Società Telespazio | March 10, 1965 | 01.898 | 2.95 |
| Ivory Coast | Télécommunications Internationales de la Côte d'Ivoire (INTELCI) | September 10, 1969 | 00.100 | – |
| Jamaica | Cable and Wireless (West Indies) Ltd. | February 4, 1969 | 00.050 | .09 |
| Japan | Kokusai Denshin Denwa Company, Ltd. | August 20, 1964 | 01.725 | 4.90 |
| Jordan | Ministry of Communications of the Hashemite Kingdom of Jordan | February 12, 1965 | 00.045 | – |
| Kenya | East African External Telecommunications Company, Ltd. | October 11, 1967 | 00.049 | .55 |
| Korea | Ministry of Communications of the Republic of Korea | February 24, 1967 | 00.049 | .53 |
| Kuwait | Ministry of Posts, Telegraphs, and Telephones of Kuwait | February 12, 1965 | 00.045 | .41 |
| Lebanon | Government of Lebanon | February 12, 1965 | 00.072 | .09 |
| Libya | Government of the Kingdom of Libya | February 12, 1965 | 00.027 | .02 |
| Liechtenstein | Government of the Principality of Liechtenstein | July 29, 1966 | 00.048 | – |
| Luxembourg | Government of Luxembourg | February 24, 1969 | 00.050 | – |
| Malaysia | Director General of the Telecommunications Department, Government of Malaysia | May 25, 1966 | 00.238 | .18 |
| Mexico | Department of Communications and Transportation of the Government of Mexico | October 25, 1966 | 01.459 | .58 |

| Country | Entity | Entry into Effect of Interim Agreement | Investment Quota | Percentage of Total Use |
|---|---|---|---|---|
| Monaco | Government of the Principality of Monaco | February 28, 1965 | 00.004 | – |
| Morocco | Government of Morocco | June 22, 1966 | 00.287 | .14 |
| The Netherlands | Government of the Kingdom of the Netherlands | August 20, 1964 | 00.863 | .60 |
| New Zealand | Postmaster General of New Zealand | February 12, 1965 | 00.404 | .12 |
| Nicaragua | Dirección General de Comunicaci- ones | February 11, 1969 | 00.050 | – |
| Nigeria | Federal Republic of Nigeria | December 8, 1965 | 00.333 | – |
| Norway | Telegrafstyret | August 31, 1964 | 00.345 | .21 |
| Pakistan | Government of Pakistan | June 30, 1965 | 00.234 | – |
| Panama | Intercontinental de Comunicaciones por Satélites, S.A. | October 20, 1967 | 00.039 | .69 |
| Peru | Empresa Nacional de Telecomunica- ciones Peru (ENTEL-Peru) | June 9, 1967 | 00.492 | .99 |
| Philippines | Philippine Com- munications Sat- ellite Corpora- tion | November 30, 1966 | 00.489 | 1.57 |
| Portugal | Administraçâo Geral dos Cor- reios, Telégrafos e Teléfones | January 14, 1965 | 00.345 | .02 |
| Saudi Arabia | Ministry of Com- munications | February 19, 1965 | 00.045 | – |
| Senegal | Société TELE- SENEGAL | May 28, 1970 | 00.050 | – |
| Singapore | Government of Singapore | June 3, 1966 | 00.096 | .41 |
| South Africa | Department of Posts and Tele- graphs of the Republic of South Africa | February 8, 1965 | 00.269 | .14 |
| Spain | Government of the State of Spain | August 20, 1964 | 00.949 | 1.80 |

| Country | Entity | Entry into Effect of Interim Agreement | Investment Quota | Percentage of Total Use |
|---|---|---|---|---|
| Sudan | Department of Posts and Telegraphs of the Government of the Republic of the Sudan | April 5, 1965 | 00.009 | – |
| Sweden | Kungl. Telestyrelsen | January 18, 1965 | 00.604 | .16 |
| Switzerland | Direction Générale des PTT | September 16, 1964 | 01.725 | 1.22 |
| Syria | Ministry of Communications of the Syrian Arab Republic | February 12, 1965 | 00.036 | – |
| Tanzania | East African External Telecommunications Company, Ltd. | June 16, 1967 | 00.049 | – |
| Thailand | Kingdom of Thailand | May 12, 1966 | 00.095 | 1.04 |
| Trinidad & Tobago | External Telecommunications (Textel) | January 20, 1970 | 00.050 | .21 |
| Tunisia | Secretariat of State for Post, Telegraph and Telephone of Tunisia | February 19, 1965 | 00.180 | – |
| Turkey | Government of the Republic of Turkey | May 6, 1968 | 00.495 | .02 |
| Uganda | East African External Telecommunications Company, Ltd. | January 5, 1968 | 00.049 | – |
| United Arab Republic | Government of the United Arab Republic | February 19, 1965 | 00.314 | – |
| United Kingdom | Post Office Corporation | August 20, 1964 | 07.245 | 9.52 |
| United States | Communications Satellite Corporation | August 20, 1964 | 52.614 | 48.11 |
| Vatican City | Government of the Vatican City State | August 20, 1964 | 00.043 | – |
| Venezuela | Ministry of Communications of Venezuela | December 30, 1965 | 00.950 | .41 |
| Viet Nam | Government of the Republic of Viet Nam | February 21, 1969 | 00.050 | .39 |

| Country | Entity | Entry into Effect of Interim Agreement | Investment Quota | Percentage of Total Use |
|---------|--------|----------------------------------------|------------------|-------------------------|
| Yemen | Yemen Arab Republic Ministry of Communications | June 29, 1965 | 00.028 | – |
| Yugoslavia | Community of the Yugoslav Posts, Telegraphs and Telephones | February 24, 1970 | 00.100 | – |
| Zambia | General Post Office, Republic of Zambia | March 20, 1970 | 00.050 | – |
| | | TOTAL | 100.00 | 100.00 |

Source: COMSAT, Annual Report to the President and the Congress, 1970, pp. 101-06.

**Notes**

# Notes

## Chapter 1
### Introduction

1. Two good readers in this area are *People, Society, and Mass Communications*, Lewis Anthony Dexter and David Manning White, eds. (New York: Free Press, 1964), and Bernard Berelson and Morris Janowitz, eds., *Reader in Public Opinion and Communication*, 2nd ed. (New York: The Free Press, 1966). See also, Bruce Lannes Smith and Chitra M. Smith, *International Communications and Political Opinion: A Guide to the Literature* (Princeton: Princeton University Press, 1956) and Charles R. Wright, *Alternative Systems of Mass Communications* (New York: Random House, 1959).

2. Examples of such studies as exist are (1) Walter S. Rogers, "International Electrical Communications," *Foreign Affairs*, 1 (December, 1922); (2) "The Political Uses of the Radio," *Geneva Studies*, 10, (Geneva, 1939), pp. 144-57; and (3) Robert D. Leigh, "The Mass-Communications Inventions and International Relations," *Technology and International Relations*, William F. Ogburn, ed. (Chicago: Chicago University Press, 1949), pp. 126-43.

3. Edward S. Corwin, *The President: Office and Powers*, 3rd ed. rev. (New York: New York University Press, 1948), p. 208.

4. John von Neumann, "Can We Survive Technology?," *Fortune*, 51 (June, 1955), p. 152.

5. William F. Ogburn, "Technology and Planning," George B. Galloway and Associates, *Planning for America* (New York: Henry, Holt, 1941).

6. Carl J. Friedrich, *Constitutional Government and Democracy* (Boston: Little, Brown, 1941), pp. 589-91.

7. Alexander Hamilton, *The Federalist,* #75.

8. Roger Hilsman, "The Foreign-Policy Consensus: An Interim Research Report," *Journal of Conflict Resolution*, 3 (December, 1959), pp. 361-82. Samuel P. Huntington, *The Common Defense* (New York: Columbia University Press, 1961). On the distinction between legislative and executive politics, see pp. 146-47. Theodore J. Lowi, "Making Democracy Safe for the World: National Politics and Foreign Policy," *Domestic Sources of Foreign Policy*, ed. James N. Rosenau (New York: Free Press, 1967), pp. 295-331.

9. James N. Rosenau, "Pre-Theories and Theories of Foreign Policy," *Approaches to Comparative and International Politics*, ed. R. Barry Farrell (Evanston: Northwestern University Press, 1966), pp. 27-92.

10. Johan Galtung, "A Structural Theory of Integration," *Journal of Peace Research*, no. 4 (1968), pp. 375-95.

11. Karl W. Deutsch, "The Impact of Science and Technology on International Politics," *Daedalus*, 88 (Fall, 1959), pp. 669-85.

12. James Madison, *The Federalist,* #10; David B. Truman, *The Governmental Process* (New York, Knopf, 1951); Robert A. Dahl, *Who Governs?* (New Haven: Yale University Press, 1961); Charles E. Lindblom, *The Intelligence of Democracy* (New York: Free Press, 1965).

13. For further development of this argument see Lindblom, *Democracy*, pp. 137-43.

14. *The Population Bomb* (New York: Ballantine, 1968).

15. *One-Dimensional Man* (Boston: Beacon Press, 1964).

16. E.g., Henry Kissinger, *The Troubled Partnership* (New York: McGraw-Hill, 1965), p. 23.

17. G. William Domhoff, *The Higher Circles: The Governing Class in America* (New York: Random House, 1970) and James Ridgeway, *The Politics of Ecology* (New York: Dutton, 1970) are examples of those who see the United States as ruled by a narrow governing class and would prefer a comprehensive approach to rational decision-making as the instrument for implementing a new order of priorities. Examples of partisans of synoptic rationality who do not see a power elite but a reactionary pluralism are Michael Harrington, *Toward A Democratic Left* (New York: Macmillan, 1968) and Theodore J. Lowi, *The End of Liberalism* (New York: W.W. Norton, 1969). It is Lowi who uses the term "interest group liberalism" to describe the current public philosophy.

Chapter 2
Traditional Communications Policy and the Coming
of the Space Age

1. Post Office Department, *Government Ownership of the Electrical Means of Communication*, Document No. 399, 63rd Cong., 2nd Sess., 1914.

2. "Roads Act," 12 Stat. 221.

3. U.S. Congress, Senate, Committee on Interstate Commerce, Subcommittee Hearings, *Cable Landing Licenses*, 66th Cong., 3rd Sess., 1921, p. 264.

4. *Foreign Relations*, 1900, p. xliii; 1901, pp. xxxiv-v; 1902, pp. xxii-xxiii. This information is reprinted in Keith Clark, *International Communications: The American Attitude* (New York: Columbia University Press, 1931), p. 153.

5. Post Office Department, *Government Control and Operation of the Telegraph, Telephone, and Marine Cable System—August 1, 1918 to July 31, 1919.* (Washington, D.C.: Government Printing Office, 1921), quoted in Keith Clark, *Communications*, p. 241.

6. Keith Clark, *Communications*, pp. 240-41.

7. James M. Herring and Gerald C. Gross, *Telecommunications* (New York: McGraw-Hill, 1936), pp. 80-83.

8. U.S. Congress, Senate, Committee on Commerce, A Subcommittee, *Hearings, Study of International Communications*, 79th Cong., 1st Sess., 1946.

9. For a short history of United States organization for propaganda, see John B. Whitton and Arthur Larson, *Propaganda: Towards Disarmament in the War of Words*, (Dobbs Ferry, New York: Oceana Publications, 1963), pp. 47-52.

10. One prominent panel whose conclusions are still relevant was the Communications Policy Board set up by President Truman in 1950. See its report, *Telecommunications—A Program for Progress* (Washington, D.C.: Government Printing Office, 1951). Also see Chapter 8 for Rostow Report.

11. "Radio Act of 1927," 44 Stat. 1162, 47 USC 4. Sec. 81. "Communications Act of 1934," 48 Stat. 1064, 47 USC 5.

12. *Treaty Series*, No. 867. 49 Stat. Part 2, 2444-2657.

13. Gabriel A. Almond, "Public Opinion and the Development of Space Technology: 1957-1960," *Outer Space in World Politics*, Joseph M. Goldsen, ed. (New York: Praeger, 1963), pp. 71-96.

14. Vernon van Dyke, *Pride and Power: The Rationale of the Space Program* (Urbana: University of Illinois Press, 1964), p. 15.

15. R. Cargill Hall, "Origins and Development of the Vanguard and Explorer Satellite Programs," *The Air Power Historian* 11 (October, 1964), pp. 101-12.

16. U.S. Congress, Senate, Committee on Armed Services, Preparedness Investigating Subcommittee, *Hearings, Inquiry into Satellites and Missile Programs*, 85th Cong., 1st & 2nd Sess., 1957-58.

17. For history of Hearings see Alison Griffith, *The National Aeronautics and Space Act: A Study of the Development of Public Policy* (Washington, D.C.: Public Affairs Press, 1962).

18. U.S. Congress, House, Committee on Government Operations, *Government Operations in Space*, House Report No. 445, 89th Cong., 1st Sess., 1965, Ch. III.

19. "National Aeronautics and Space Act of 1958," 72 Stat. 426; 42 U.S.C. 2451, Title I, Sec. 102.

20. "Communications Act of 1934," Sections 303, 212. Congressman William Fitts Ryan took this view in 1961. See U.S. Congress, House, Committee on Science and Astronautics, *Hearings, Communications Satellites*, Part 1, 87th Cong., 1st Sess., 1961, p. 543.

21. "Space Act of 1958," Section 103 (1).

## Chapter 3
## Early Developments in Space Communications

1. Testimony of Fred C. Alexander of the Office of Civil and Defense Mobilization, U.S. Congress, House, Committee on Science and Astronautics, *Hearings, Communications Satellites, Part 2*, 87th Cong., 1st Sess., 1961, p. 602, hereinafter cited as House Committee on Science and Astronautics, *Communications Satellites Hearings* (in two parts).

2. Report of the Ad Hoc Committee on the Peaceful Uses of Outer Space, July 14, 1959. Reprinted as Appendix A of U.S. Congress, Senate, Committee on Aeronautical and Space Sciences, *Radio Frequency Control in Space Telecommunications*, Committee Print, 86th Cong., 2nd Sess., 1960, p. 123, hereinafter cited as Senate Committee on Aeronautical and Space Sciences, *Radio Frequency Control*.

3. Senate Committee on Aeronautical and Space Sciences, *Radio Frequency Control*, p. 34.

4. Ibid., p. 32.

5. Ibid., p. 33.

6. Ibid., p. 35.

7. FCC, Docket 11866, Memorandum, Opinion and Order, Released October 5, 1960, reprinted in U.S. Congress, Senate, Committee on Aeronautical and Space Sciences, *Policy Planning for Space Telecommunications*, Staff Report, 86th Cong., 2nd Sess., 1960, p. 180, hereinafter cited as Senate Committee on Aeronautical and Space Sciences, *Policy Planning Report*.

8. Ibid., p. 188.

9. Senate Committee on Aeronautical and Space Sciences, *Policy Planning Report*, p. 78. This report had been prepared by Dr. Edward Wenk, then science advisor in the Legislative Reference Service of the Library of Congress.

10. Ibid., p. 15.

11. U.S. Congress, House, Committee on Science and Astronautics, *Commercial Applications of Space Communications Systems*, Report No. 1279, 86th Cong., 2nd Sess., 1961, p. 20.

12. Senate Committee on Aeronautical and Space Sciences, *Policy Planning Report*, p. 78.

13. Ibid., p. 123.

14. Ibid., pp. 124-25.

15. Ibid., pp. 120-26.

16. Ibid., p. 126.

17. *New York Times*, October 13, 1960, pp. 1, 20.

18. *New York Times*, December 31, 1960.

19. U.S. Congress, Senate, Committee on Commerce, Communications Subcommittee, *Space Communications and Allocation of Radio Spectrum*, 87th Cong., 1st Sess., 1961, p. 106.

20. House Committee on Science and Astronautics, *Communications Satellite Hearings*, Part 1, p. 464.

21. *Department of State Bulletin*, 44 (February 13, 1961), p. 213.

22. Report to the President-Elect of the Ad Hoc Committee on Space, Reprinted in U.S. Congress, House; Committee on Science and Astronautics, *Defense Space Interests*, 87th Cong. 1st Sess., 1961, pp. 17-18. This listing is abbreviated.

23. Ibid., pp. 22-23.

24. U.S. Congress, House, Committee on Science and Astronautics and Subcommittees 1, 3, and 4. *Hearings, 1962 NASA Authorization*, Part 2, 87th Cong., 1st Sess., 1961, pp. 726-30. The entire NASA Research and Development budget request for communications was $44.6 million.

25. Reprinted in House Committee on Science and Astronautics, *Communications Satellites Hearings*, Part 1, pp. 499-500.

26. FCC, Docket No. 14024.

27. Ibid., Statement of the Department of Justice, May 5, 1961. Reprinted in House Committee on Science and Astronautics, *Communications Satellites Hearings*, Part 1, p. 561.

28. Docket No. 14024, First Report, May 24, 1961. Reprinted in ibid., pp. 539-41.

29. Docket No. 14024, Proceedings, June 5, 1961. Reprinted in ibid., pp. 563-81.

30. Ibid., pp. 567-69.

31. Ibid., p. 578.

32. Ibid., p. 576.

33. Letter from the President to the Vice President, June 15, 1961. Reprinted in U.S. Congress, Senate, Select Committee on Small Business, Subcommittee on Monopoly, *Hearings, Space Satellite Communications*, 86th Cong., 1st Sess., 1961, Part 1, p. 511, hereinafter cited as Senate Subcommittee on Monopoly, *Space Satellite Communications Hearings* (in two parts).

34. *Department of State Bulletin*, 44 (June 12, 1961), pp. 903-10.

35. U.S. Congress, House, Committee on Science and Astronautics, *1962 NASA Authorization*, Part 3, 87th Cong., 1st Sess., 1961, p. 1039.

36. House Committee on Science and Astronautics, *Communications Satellites Hearings*, Part 2, p. 761.

37. Reprinted in Senate Subcommittee on Monopoly, *Space Satellite Communications Hearings*, p. 16.

38. House Committee on Science and Astronautics, *Communications Satellites Hearings*.

39. U.S. Congress, House, Committee on Interstate and Foreign Commerce, *Hearings, Communications Satellites*, 87th Cong., 1st Sess., 1961, hereinafter cited as House Committee on Interstate and Foreign Commerce, *Communications Satellites Hearings*.

40. U.S. Congress, House, Committee on the Judiciary, Antitrust Subcommittee, *Hearings, Antitrust Consent Decrees and the Television Broadcast Industry*, 87th Cong., 1st Sess., 1961.

41. Senate Subcommittee on Monopoly, *Space Satellite Communications Hearings*.

42. Dr. John R. Pierce, "Orbital Radio Relays," *Jet Propulsion*, 25 (April, 1955).

43. House Committee on Science and Astronautics, *Communications Satellites Hearings*, Part 1, p. 12.

44. Ibid., p. 408.

45. Ibid., p. 15.

46. U.S. Congress, House, Committee on Science and Astronautics, *Hearings, Satellites for World Communications*, 86th Cong., 1st Sess., 1959, p. 56, hereinafter cited as House Committee on Science and Astronautics, *Satellites for World Communications Hearings*.

47. House Committee on Science and Astronautics, *Communications Satellites Hearings*, Part 1, p. 409.

48. Ibid., pp. 394-98.

49. Senate Subcommittee on Monopoly, *Space Satellite Communications Hearings*, Part 1, p. 379. The linkage of the free world and the U.N. bears note. President Kennedy envisaged the system serving the "globe."

50. House Committee on Science and Astronautics, *Satellites for World Communications Hearings*, p. 107.

51. Senate Subcommittee on Monopoly *Space Satellite Communications Hearings*, p. 251.

52. Ibid., p. 254.

53. Ibid., pp. 175-76.

54. Ibid., p. 198.

55. Ibid., p. 200. AT&T buys all its equipment from its affiliate, Western Electric.

56. Ibid., p. 642.

57. Ibid., p. 30.

58. House Committee on Science and Astronautics, *Communications Satellites Hearings*, Part 1, p. 136.

59. Senate Subcommittee on Monopoly, *Space Satellite Communications Hearings*, Part 1, p. 31. For similar views, see the testimony of the Assistant Attorney General, Nicholas B. Katzenbach, House Committee on Science and Astronautics, *Communications Satellites Hearings*, Part 2, p. 717.

60. Senate Subcommittee on Monopoly, *Space Satellite Communications Hearings*, pp. 38-39, 48. Similar fears were expressed by Bendix, Dunn Engineering, and the Midwest Technical Development Corporation. See pp. 411, 420, and 422 respectively.

61. House Committee on Science and Astronautics, *Communications Satellites Hearings*, p. 769.

62. Senate Subcommittee on Monopoly, *Space Satellite Communications Hearings*, p. 271.

63. Ibid., p. 283.

64. House Committee on Science and Astronautics, *Communications Satellites Hearings*, Part 1, p. 659.

65. Ibid., pp. 658-59.

66. Ibid., p. 659.

67. Ibid., p. 661.

68. Ibid., Part 2, pp. 704-15.

69. Ibid., p. 714.

70. Other Executive participants offered their views at the hearings. The USIA was principally interested in space communications as a means of giving "the peoples of the world, and particularly the emerging nations, an opportunity to make free choice on the basis of maximum information." (See testimony, ibid., pp. 583-600.) The views of the Office of Civil and Defense Mobilization related to its responsibilities for frequency management and coordination of Government telecommunications policies. (See testimony, ibid., pp. 600-33.) The responsibilities of the National Aeronautics and Space Council included advice to the President on overall policy. (See testimony, ibid., pp. 760-61.) The Federal Aviation Agency was interested in space communications as a means of extending air traffic control and providing increased communications coverage for the International Civil Aviation Organization area control centers. (See testimony, Senate Subcommittee on Monopoly, *Space Satellite Communications Hearings*, Part 1, pp. 357-76.) The Small Business Administration was interested in providing safeguards for small business in the proposed communications satellite system. (See testimony, ibid., pp. 306-18.)

71. Letter to the President from certain members of Congress, August 24, 1961. Reprinted in U.S. Congress, Senate, Committee on Foreign Relations, *Hearings, Communications Satellites Act of 1962*, 87th Cong., 2nd Sess., 1962, pp. 51-54.

72. Ibid., p. 29.

73. Senate Subcommittee on Monopoly, *Space Satellite Communications Hearings*, pp. 397, 413.

74. Ibid., p. 215.

75. Ibid., pp. 405-06.

76. Ibid., pp. 398-409.

77. Ibid., pp. 381, 386.

78. House Concurrent Resolution 360, 87th Cong., 1st Sess., 1961. Reprinted ibid., p. 409.

79. Ibid., p. 381.

80. The views of Lockheed, GT&E, AT&T, and ITT on the foreign challenge may be found in ibid., at pp. 200, 166, 250, and 203 respectively.

81. House Committee on Science and Astronautics, *Communications Satellites Hearings*, p. 5.

82. Senate Subcommittee on Monopoly, *Space Satellite Communications Hearings*, pp. 121-23.

83. Ibid., p. 268.

84. Ibid., p. 128.

85. Ibid., p. 129.

86. Ibid., p. 136.

87. Ibid., p. 161.

88. Ibid., p. 164.

89. Ibid., p. 167.

90. Ibid., Part 2, p. 658.

91. House Committee on Science and Astronautics, *Communications Satellites Hearings*, Part 2, p. 888.

92. U.S. Congress, Senate, Committee on Aeronautical and Space Sciences, *Communications Satellites: Technical, Economic, and International Developments*, Staff Report, 87th Cong., 2nd Sess., 1962, p. 196.

93. Docket No. 14024, Comments by Hughes Aircraft Corporation November 13, 1961. Reprinted in Senate Subcommittee on Monopoly, *Space Satellite Communications Hearings*, Part 2, pp. 734-37.

94. Senate Subcommittee on Monopoly, *Space Satellite Communications Hearings*, Part 1, p. 252.

95. Ibid., p. 205.

96. Ibid., p. 380. For the views of General Electric and Bendix see ibid., pp. 390 and 41 respectively.

97. Ibid., p. 425.

98. Ibid., p. 451.

99. Senate Committee on Aeronautical and Space Sciences, *Radio Frequency Control*, p. 12.

100. Ibid., p. 43.

101. For a provocative analysis along these lines see Marshall McLuban, *Understanding Media: The Extensions of Man* (New York: McGraw Hill Paperbacks, 1965).

102. For both views see House Committee on Science and Astronautics, *Commercial Applications of Space Communications Systems*, Report No. 1279, 86th Cong., 2nd Sess., 1961, pp. 12, 21.

103. House Committee on Science and Astronautics, *Communications Satellites Hearings*, Part 1, p. 134.

104. Ibid., p. 306. These are FCC figures.

105. Docket No. 14024, *Report of the Ad Hoc Carrier Committee*, October 12, 1961, p. 15. Reprinted in Senate Subcommittee on Monopoly, *Space Satellite Communications Hearings*, Part 2, Appendix I (a).

106. Communications Policy Board, *Telecommunications*, pp. 174-76.

107. Senate Committee on Aeronautical and Space Sciences, *Policy Planning Report*, p. 99.

108. Docket No. 14024, *Report of the Ad Hoc Carrier Committee*.

109. Statement of Ambassador Stevenson, Committee I of the U.N., December 4, 1961. Reprinted in Senate Committee on Aeronautical and Space Sciences, *Radio Frequency Control*, pp. 146-51.

110. Senate Subcommittee on Monopoly, *Space Satellite Communications Hearings*, Part 2, p. 633.

111. Ibid., pp. 580, 557, et passim.

112. Ibid., pp. 574-83.

113. Docket No. 14024, *Report of the Ad Hoc Carrier Committee*, p. 44.

114. Senate Subcommittee on Monopoly, *Space Satellite Communications Hearings*, p. 547.

115. Ibid., p. 567.

116. *Department of State Bulletin*, 45 (October 16, 1961), p. 622.

117. Statement of Ambassador Stevenson, Senate Committee on Aeronautical and Space Sciences, *Radio Frequency Control*, p. 146.

118. Ibid., pp. 149-50. The other parts were: (1) the establishment of a regime of law and order in outer space, (2) the encouragement of open and orderly conduct of outer space activities, (3) a worldwide weather system for research and prediction, and (4) the revitalization of the U.N. Committee on the Peaceful Uses of Outer Space.

119. The term "penetrated system" comes from James N. Rosenau, "Pre-Theories and Theories of Foreign Policy," *Approaches to Comparative and International Politics*, ed. R. Barry Farrell (Evanston: Northwestern University Press, 1966), pp. 27-92.

120. Analogous with the tariff issue: See Raymond A. Bauer, Ithiel de Sola Pool, and Lewis Anthony Dexter, *American Business and Public Policy: The Politics of Foreign Trade* (New York: Atherton Press, 1963).


Chapter 4
The Passage of the Communications Satellite Act of 1962

1. Interview, Dr. Edward Welsh, October 12, 1965. For idea that national pride as well as national prestige at issue in space policy see Vernon van Dyke, *Pride and Power: The Rationale of the Space Program* (Urbana: University of Illinois Press, 1964), p. 15.

2. For the development of this distinction, see U.S. Congress, Senate, Committee on Government Operations, Subcommittee on National Security and International Operations, *Hearings, Conduct of National Security*, 89th Cong., 1st Sess., 1965.

3. U.S. Congress, Senate, Committee on the Judiciary, Antitrust and Monopoly Subcommittee, *Hearings, Antitrust Problems of the Space Satellite Communications System*, Part 1, 87th Cong., 2nd Sess., 1962, p. 117, hereinafter cited as Senate Antitrust and Monopoly Subcommittee, *Satellite Communications Hearings* (in two parts). But Mr. Loevinger considered this to be a practical compromise in order to get the job done quickly. Ordinarily such an arrangement would be seen as an interlocking directive contrary to the Clayton Act; ibid., pp. 143, 147, 149.

4. House Committee on Interstate and Foreign Commerce, *Communications Satellites Hearings*, Part 2, p. 404.

5. Ibid., pp. 403-04.

6. Ibid.

7. A RAND study for NASA at this time elaborated on the inefficiencies of the FCC regulation. See Leland Johnson, "Communications Satellites and Telegraph Rates: Problems of Governmental Regulation," RM-2845, October, 1961. Also see excerpts of the organization and management survey of the FCC done by Booz, Allen, and Hamilton for the Bureau of the Budget. Senate Antitrust and Monopoly Subcommittee, *Satellite Communications Hearings*, Part 2, pp. 653-67.

8. U.S. Congress, Senate, Committee on Aeronautical and Space Sciences, *Hearings, Communications Satellite Legislation*, 87th Cong., 2nd Sess., 1962, pp. 264, 304, 424, hereinafter cited as Senate Committee on Aeronautical and Space Sciences, *Communications Satellite Legislation Hearings*. See also pp. 19-20, 274.

9. House Committee on Interstate and Foreign Commerce, *Communications Satellites Hearings*, Part 2, p. 360.

10. Ibid., p. 569. See also pp. 395-96.

11. Senate Antitrust and Monopoly Subcommittee, *Satellite Communications Hearings*, Part 1, p. 3.

12. Senate Committee on Aeronautical and Space Sciences, *Communications Satellite Legislation Hearings*, pp. 353, 364-65.

13. U.S. Congress, Senate, Committee on Commerce, Hearings, *Communications Satellite Legislation*, 87th Cong., 2nd Sess., 1962, p. 311, hereinafter cited as Senate Commerce Committee, *Communications Satellites Hearings*.

14. For similar views see testimony of Representative Emanuel Celler, House Committee on Interstate and Foreign Commerce, *Communications Satellites Hearings*, Part 2, pp. 598-99.

15. Senate Commerce Committee, *Communications Satellites Hearings*, p. 369.

16. Ibid., pp. 268-69.

17. For example, Senate Committee on Aeronautical and Space Sciences, *Communications Satellite Legislation Hearings*, p. 361.

18. Ibid., pp. 429, 437.

19. Senate Commerce Committee, *Communications Satellites Hearings*, p. 310.

20. Ibid., p. 311.

21. Senate Committee on Aeronautical and Space Sciences, *Communications Satellite Legislation Hearings*, p. 356.

22. Ibid.

23. Ibid., p. 346. Points 2 and 3 could be construed as responsibilities the Government should perform in its conduct of foreign relations. Points 1 and 4 would be powers whose exercise would affect both domestic and foreign interests. This listing is paraphrased.

24. Senate Commerce Committee, *Communications Satellites Hearings*, p. 341.

25. Ibid., p. 339.

26. Ibid., p. 334.

27. Ibid., p. 275.

28. Senate Committee on Aeronautical and Space Sciences, *Communications Satellite Legislation Hearings*, p. 161.

29. Ibid., p. 165.

30. Ibid.

31. Ibid., p. 171.

32. See Senate Antitrust and Monopoly Subcommittee, *Satellite Communications Hearings*, Part 1, p. 111.

33. Senate Commerce Committee, *Communications Satellites Hearings*, p. 292.

34. Ibid., p. 293.

35. House Committee on Interstate and Foreign Commerce, *Communications Satellites Hearings*, Part 2, p. 563.

36. Senate Antitrust and Monopoly Subcommittee, *Satellite Communications Hearings*, Part 1, p. 48.

37. House Committee on Interstate and Foreign Commerce, *Communications Satellites Hearings*, Part 2, p. 385.

38. Ibid., p. 362.

39. Ibid., p. 399, 405.

40. Ibid., p. 406.

41. Senate Committee on Aeronautical and Space Sciences, *Communications Satellite Legislation Hearings*, p. 19.

42. AID and its predecessor agencies had obligations for foreign telecommunications for FY 1955 through FY 1961 of $94 million, House Committee on Interstate and Foreign Commerce, *Communications Satellites Hearings*, Part 2, p. 481.

43. Senate Committee on Aeronautical and Space Sciences, *Communications Satellite Legislation Hearings*, p. 481.

44. Ibid., pp. 21-23.

45. Ibid., p. 68.

46. Ibid., pp. 14-16.

47. Ibid., pp. 260-61.

48. Ibid., p. 266.

49. Ibid., p. 281.

50. In addition to AT&T, there were Hawaiian Telephone, ITT, Communications Workers of America, and the U.S. Independent Telephone Association. The Communications Workers favored this solution because they could get better labor contracts from private industry than from the Government. Ibid., p. 444.

51. E.g., see ibid., pp. 214-15, 227, 309-32.

52. Senate Antitrust and Monopoly Subcommittee, *Satellite Communications Hearings*, Part 2, p. 390.

53. Ibid.

54. Ibid., p. 437. This is one of the few references in all the Congressional hearings to the attitudes of the less-developed countries.

55. U.S. Congress, Senate Committee on Aeronautical and Space Sciences, *Communications Satellite Act of 1962*, S. Rept. 1319 to Accompany S. 2814, pp. 3-4.

56. For the original bill, S. 2814, and the Space Committee's Amendments, see Senate Commerce Committee, *Communications Satellites Hearings*, pp. 4-10.

57. Senate Committee on Aeronautical and Space Sciences, *Communications Satellite Legislation Hearings*, pp. 304.

58. *Report of the Committee on Aeronautical and Space Sciences*, p. 5.

59. U.S. Congress, House, Committee on Interstate and Foreign Commerce, 87th Cong., 2nd Sess., 1962, H. Rept. 1636 to accompany H.R. 11040.

60. U.S. Congress, Senate, Committee on Commerce, *Communications Satellite Act of 1962*, 87th Cong., 2nd Sess., 1962, S. Rept. 1584 to accompany H.R. 11040.

61. Ibid., p. 12, 102 (c).

62. Ibid., p. 15, Sec. 201 (a) (4).

63. Ibid., p. 23-24.

64. Testimony of USIA, U.S. Congress, Senate, Committee on Foreign Relations, Hearings, *Communications Satellite Act of 1962*, 87th Cong., 2nd Sess., 1962, p. 126.

65. Quotes from USIA, see ibid.

66. Ibid., p. 45.

67. Ibid., p. 175.

68. U.S. Congress, Senate, Committee on Foreign Relations, 87th Cong., 2nd Sess., 1962, *Communications Satellite Act of 1962*, S. Rept. 1873 to accompany H.R. 11040. Separate views of Senator Frank Church of Idaho who introduced an amendment to Sec. 102 (d) may be found on pp. 14-15.

69. 76 Stat. 419, 47 U.S.C. § § 701-44; *Congressional Quarterly*, 20 No. 35, p. 1503.

70. John von Neumann, "Can We Survive Technology," *Fortune*, 51 (June, 1955), p. 152.

71. Senate Committee on Aeronautical and Space Sciences, *Communications Satellite Legislation Hearings*, p. 144.

72. U.S. Bureau of the Census, *Statistical Abstract of the United States: 1963*, 84th ed. (Washington, D.C.: 1963), p. 515.

73. *Annual Report of the FCC, FY 1964* (Washington, D.C.: 1963), pp. 126-30.

74. Bruce M. Russet et al., *World Handbook of Political and Social Indicators* (New Haven: Yale University Press, 1964), Tables 31-38.

Chapter 5
The Interim Arrangements for a Global Commercial
Communications Satellite System

1. A list of delegates may be found in *U.S. Congressional Record*, 88th Cong., 2nd Sess., January 9, 1964, 110 (daily ed.), pp. 166-74, hereinafter cited as *U.S. Congressional Record*, 88th Cong., 2nd Sess.

2. Leonard Jaffe (NASA), T. Arthur Smith (Operations Research, Inc.), L.O. Attaway (Rand), "Communications," Vol. 12 of U.S. Papers prepared for the UN Conference on the Application of Science and Technology for the Benefit of the Less-developed Areas, February, 1963, p. 197. The "realism" of their assessment contrasts with the symbolic importance of the meaning of communications satellites for the less developed countries a year earlier in the Congressional debates. For a contrary view to Jaffe, see Senate Committee on Aeronautical and Space Sciences, *Policy Planning Report*, p. 99.

3. Report of the Chairman of the U.S. Delegation. Reprinted in *U.S. Congressional Record*, 88th Cong., 2nd Sess., pp. 166-74, 168-72.

4. Ibid., pp. 168-69.

5. Ibid., p. 170.

6. 2,800 mc/s were allocated for Region 1 (Europe, the USSR, and Africa), 2,600 for Region 2 (the Western Hemisphere), and 2,675 for Region 3 (Asia, except the USSR), ibid.

7. Quoted in U.S. Congress, House, Committee on Science and Astronautics, Subcommittee on Space Sciences and Applications, *Hearings, NASA Authorization, 1965*, 88th Cong., 2nd Sess., 1964, p. 172.

8. Ibid., p. 172.

9. Ibid., p. 165.

10. Ibid., p. 175.

11. U.S. Congress, Senate, Committee on Aeronautical and Space Sciences, *Communications Satellites: Technical, Economic, and International Developments*, Staff Report, 87th Cong., 2nd Sess., 1962, p. 126. Dr. MacQuivey wrote the Report. He was then in the Telecommunications Division of the Department of State.

12. Samuel P. Huntington, *The Common Defense* (New York: Columbia University Press, 1961), p. 146.

13. The Communications Satellite Act, Sections 201(a)(5) and 305(a)(1).

14. The Communications Satellite Act, Section 301.

15. Ibid., Section 302.

16. The biographies of these men may be found in U.S. Congress, Senate, Committee on Commerce, *Communications Satellite Incorporation, Hearings*, 88th Cong., 1st Sess., 1963, pp. 2-16, hereinafter cited as Senate Committee on Commerce, *Communications Satellite Incorporation Hearings*.

17. At the same time, Dr. Joseph V. Charyk, formerly Under Secretary of the Air Force, businessman, and university scientist was appointed President and principal operating officer. Also, in February, Comsat's articles of incorporation were approved. The articles and bylaws may be found in Senate Committee on Commerce, *Communications Satellite Incorporation Hearings*, pp. 35-39.

18. Letter of Leo D. Welch to E. William Henry, Chairman of FCC, August 7, 1963. Reprinted as Attachment C in U.S. Congress, House, Committee on Interstate and Foreign Commerce, *Communications Satellite Act of 1962–The First Year*, H. Rept. 809, 88th Cong., 1st Sess., 1963, p. 19, hereinafter cited as House Committee on Interstate and Foreign Commerce, *Communications Satellite Report*.

19. In 1963 the FCC had permitted the Corporation "to borrow $1,900,000 pursuant to a line of credit agreement for $5 million entered into by the corporation with 10 commercial banks." House Commitee on Interstate and Foreign Commerce, *Communications Satellite Report*, p. 11.

20. House Committee on Interstate and Foreign Commerce, *Communications Satellite Report*, p. 29.

21. Comsat, Statement Relating to Anticipated Participation by Foreign Governments or Business Entities in Such a System, Reprinted ibid., p. 6.

22. Richard N. Gardner, "Space Meteorology and Communications: A Challenge to Science and Diplomacy," Department of State Bulletin, May 13, 1963, p. 774. It may be asked why competition is beneficial within the United States and disastrous outside.

23. Ibid.

24. Ibid.

25. (1) U.S. Congress, Senate, Committee on Commerce, *Hearing, Communications Satellite Incorporators*, 88th Cong., 2nd Sess., 1963, hereinafter cited as Senate Commerce Committee, *Incorporation Hearing*.

(2) U.S. Congress, Senate, Committee on Aeronautical and Space Sciences, *Hearing, Nomination of Incorporators*, 88th Cong., 2nd Sess., 1963, hereinafter cited as Senate Aeronautical and Space Sciences Committee, *Incorporation Hearing*.

26. The views of Senators Monroney, Pastore, Hart, and Magnuson may be found in Senate Commerce Committee, *Incorporation Hearing*, pp. 68-70, 80-82. For the views of Senators Symington, Anderson, and Keating, see Senate Aeronautical and Space Sciences Committee, *Incorporation Hearing*, pp. 70-73, 77, 91-93.

27. Senate Commerce Committee, *Incorporation Hearing*, pp. 82-83.

28. Ibid., p. 66.

29. Ibid., p. 28.

30. Senate Commerce Committee, *Incorporation Hearing*, p. 86.

31. Ibid., p. 70.

32. Senate Aeronautical and Space Sciences Committee, *Incorporation Hearing*, p. 91.

33. U.S. Congress, House, *NASA Authorization for Fiscal Year 1964, Conference Report*, Rept. 706, 88th Cong., 1st Sess., 1963, pp. 1-2.

34. See Senator Monroney's distress at Government involvement, U.S. Congress, Senate, Committee on Commerce, Subcommittee on Communications, *Hearings, Satellite Communications*, 88th Cong., 1st Sess., 1963, p. 60.

35. Senate Commerce Committee, *Incorporation Hearing*, p. 85.

36. The following information on ESRO, ELDO, and EUROSPACE is drawn from British Information Services, *Britain and Space Research* (London: Central Office of Information, 1965). See also, U.S. Congress, Senate, Committee on Aeronautical and Space Sciences, *International Cooperation and Organization for Outer Space*, Staff Report, Document 56, 89th Cong., 1st Sess., 1965, pp. 105-17; 123-28.

37. British Information Services, *Britain and Commonwealth Telecommunications* (London: Central Office of Information, 1963), p. 26.

38. Department of State, *Summary of Activities of Department of State Relating to the Communications Satellite Act of 1962* (September 20, 1963). Reprinted in House Committee on Interstate and Foreign Commerce, *Communications Satellite Report*, pp. 25-27.

39. Ibid.

40. U.S. Congress, House, Committee on Government Operations, Military Operations Subcommittee, *Hearings, Satellite Communications–1964*, Part 2, 88th Cong., 2nd Sess., 1964, p. 661, hereinafter cited as House Military Operations Subcommittee, *Satellite Communications Hearings*.

41. Department of State, "Summary of European Regional Organization in the Communications Satellite Field," Reprinted in *U.S. Congressional Record*, 88th Cong., 2nd Sess., p. 175.

42. Department of State, *U.S. Congressional Record*, 88th Cong., 2nd Sess.

43. Ibid.

44. House Military Operations Subcommittee, *Satellite Communications Hearings*, Part 1, p. 316.

45. Ibid., Part 2, p. 663.

46. Ibid.

47. Great Britain, 61 *Parliamentary Debates* (Commons), 690 (1964), p. 421. As quoted from House Military Operations Subcommittee, *Satellite Communications Hearings*, Part 1, p. 28.

48. See testimony of Abram Chayes, House Military Operations Subcommittee, *Satellite Communications Hearings*, Part 1, p. 363.

49. Ibid., p. 661.

50. Ibid., p. 664.

51. See Article IX of the Interim Arrangements. 2 U.S.T. 1705, T.I.A.S. 5646.

52. See Annex of Special Agreement, in Interim Arrangements.

53. Carter, House Military Operations Subcommittee, *Satellite Communications Hearings*, Part 2, p. 667.

54. Article V (e) of the Interim Arrangements.

55. Carter, House Military Operations Subcommittee, *Satellite Communications Hearings*, Part 2, p. 681.

56. House Military Operations Subcommittee, *Satellite Communications Report*, p. 100.

57. House Military Operations Subcommittee, *Satellite Communications Hearings*, Part 2, p. 739.

58. Subsequently, an arbitration agreement was opened for signature June 4, 1965 and entered into force on November 21, 1966. *Department of State Bulletin*, December 12, 1966, p. 906.

59. Department of State, Transcript of Press Briefing, July 24, 1964.

60. Here, one is reminded of Don K. Price's three propositions concerning the effects of the scientific revolution: (1) The scientific revolution is moving the public and private sectors closer together; (2) the scientific revolution is bringing a new order of complexity into the administration of public affairs; and (3) the scientific revolution is upsetting our system of checks and balances. *The Scientific Estate* (Cambridge: Belknap Press, 1965), pp. 15-16.

**Chapter 6**
**Space Communications and National Security**

1. For a more radical analysis of military satellite communications policy than I provide in this chapter see Herbert I. Schiller, *Mass Communications and American Empire* (Boston: Beacon Press, 1971).

2. House Military Operations Subcommittee, *Satellite Communications Report*, pp. 117-18.

3. For a history of Project ADVENT, see ibid., pp. 123-58.

4. House Military Operations Subcommittee, *Satellite Communications Hearings*, Part 1, pp. 10-11.

5. Ibid., p. 4.

6. Memorandum, "Establishment of the National Communications System," in ibid., Part 2, Appendix 4B.

7. Memorandum, ibid. The major operation agencies of the NCS are Defense, which accounts for about 75 percent of all Government-owned or leased circuitry, the Federal Aviation Agency, State, NASA, the General Services Administration, Atomic Energy Commission, Commerce, Interior, USIA, and the FCC.

8. House Military Operations Subcommittee, *Satellite Communications Report*, p. 83.

9. House Military Operations Subcommittee, *Satellite Communications Hearings*, Part 1, p. 5.

10. House Military Operations Subcommittee, *Satellite Communications Report*, p. 123.

11. U.S. Congress, House, Committee on Science and Astronautics, *Hearings, Satellites for World Communications*, 86th Cong., 1st Sess., 1959, p. 19.

12. House Committee on Science and Astronautics, *Communications Satellites Hearings*, Part 2, p. 697.

13. House Military Operations Subcommittee, *Satellite Communications Hearings*, Part 1, p. 297.

14. Ibid., p. 279.

15. Ibid., p. 27.

16. Ibid., p. 188.

17. Ibid., p. 204.

18. Ibid., p. 14-15, 304, 417.

19. Ibid., pp. 422, 440.

20. Ibid., pp. 421, 426.

21. Ibid., p. 420.

22. Ibid., p. 75.

23. Ibid., p. 448. This was in addition to the $313 million already invested by the Government in the Philco contract.

24. Ibid., p. 303. The phrase "agree first and figure later" was used by Mr. Roback of the Committee Staff to suggest the pattern of policy-making. Mr. Clark said, "That may describe it."

25. Ibid., p. 305. Full testimony pp. 291-307.

26. Ibid., pp. 425-428, et passim.

27. Ibid., p. 428.

28. Ibid., p. 75, et passim.

29. Ibid., pp. 317-18. See Section 102 (d) of the Communications Satellite Act.

30. Ibid., p. 487.

31. House Military Operations Subcommittee, *Satellite Communications Hearings*, Part 1, pp. 259, 264, 266, 270.

32. Ibid., Part 2, p. 745.

33. Ibid., p. 756.

34. Letter of Llewellyn E. Thompson to Hon. James D. O'Connell, July 10, 1964. Reprinted in ibid., p. 757.

35. Ibid., p. 675.

36. Ibid., p. 696.

37. Ibid., pp. 676, 703.

38. House Military Operations Subcommittee, *Satellite Communications Hearings*, p. 32.

39. I say apparently both because the figures were estimates and estimates tend to vary depending who makes them and what variables are included or excluded. For arguments against the new program involving the reliability of the Titan III-C as compared with the Atlas Agena, see ibid., pp. 705-17.

40. U.S. Congress, Senate, Committee on Aeronautical and Space Sciences, *Hearings, NASA Authorization for Fiscal Year 1966*, 89th Cong., 1st Sess., 1965, p. 612.

41. Ibid., p. 610.

42. U.S. Congress, House, Committee on Government Operations, Military Operations Subcommittee, *Hearings, Military Communications–1968*, 90th Congress, 2nd Sess., 1968, p. 25.

43. *Aviation Week and Space Technology*, June 24, 1968, p. 26; January 11, 1971, p. 40; and April 10, 1972, p. 9.

44. U.S. Congress, House, Committee on Government Operations, Military Operations, Subcommittee, *Hearings, Missile and Space Ground Support Operations*, 89th Cong., 2nd Sess., 1965, hereinafter cited as House Military Operations Subcommittee, *Missile and Space Ground Support Hearing*.

45. U.S. Congress, Senate, Committee on Aeronautical and Space Sciences, *Hearings, National Communications Satellite Programs*, 89th Cong., 2nd Sess., 1966, hereinafter cited as Senate Committee on Aeronautical and Space Sciences, *National Communications Satellite Hearings.*

46. House Military Operations Subcommittee, *Missile and Space Ground Support Hearing*, p. 29.

47. Ibid., p. 35.

48. Ibid., p. 150.

49. Ibid., pp. 26-27, et passim.

50. Ibid., p. 77.

51. House Military Operations Subcommittee, *Missile and Space Ground Support Hearing*, pp. 150-51, et passim. The contract as finally negotiated in early 1966 calls for a three year contract totaling over $38.9 million.

52. U.S. Congress, House, Committee on Government Operations, Military Operations Subcommittee, *Hearings, Government Use of Satellite Communications*, 89th Congress, 2nd Sess., 1966, pp. 44-45.

53. Ibid., pp. 703-05.

54. Ibid., pp. 45, 81.

55. Reprinted in ibid., pp. 706-18.

56. Ibid., p. 717.

57. U.S. Congress, House, *Report No. 1836*, 89th Congress, 2nd Sess., 1966, p. 7.

58. U.S. Congress, House, *Report No. 613*, 90th Congress, 1st Sess., 1967, p. 10.

## Chapter 7
## INTELSAT and INTERSPUTNIK:
### American-Soviet Relations

1. It is unlikely that the Soviets would cooperate anyway, as their share of international telecommunications traffic is less than 1 percent. Therefore, if they

should join INTELSAT, they might not even be represented on the Board of Governors where the United States would have 40 percent of the vote. However, a political rather than a traffic formula might encourage Soviet participation or at least growing cooperation with INTELSAT. For suggestions, see Thomas L. Shillinglaw, "The Soviet Union and International Satellite Telecommunications," *Stanford Journal of International Studies*, 5 (June, 1970), pp. 199-226.

2. For tight, logical distinctions, see Anatol Rapaport, *Fights, Games, and Debates* (Ann Arbor: University of Michigan Press, 1960).

3. *The New York Times*, December 6, 1957, p. 33.

4. As quoted in Dallas W. Smythe, "Communications Satellites," *Bulletin of the Atomic Scientists*, 17 (February, 1961), p. 67.

5. "Global Television Programs," *Aviation Week and Space Technology*, August 8, 1960, p. 21.

6. As quoted by Senator Warren G. Magnuson, "Planning for Space Communications," *Astronautics*, 7 (January, 1962), p. 19.

7. Charles S. Sheldon II, "The Challenge of International Competition," in U.S. Congress, Senate, Committee on Aeronautical and Space Sciences, *International Cooperation and Organization for Outer Space*, Staff Report, 89th Cong., 1st Sess., 1965, pp. 427-77, 454.

8. Ibid.

9. *Aviation Week and Space Technology*, August 19, 1968, p. 20.

10. T.I.A.S., No. 6347.

11. T.I.A.S., No. 6599.

12. Res. 1472 (14), 1959; Res. 1721 (16), 1961; Res. 1802 (17), 1962; Res. 1962 (18), 1963; Res. 2345 (22), 1967.

13. United Nations Document A/AC.105/OR.2, March 29, 1962, pp. 22-25.

14. United Nations Document A/AC.105/PV.4, March 21, 1962, pp. 2-13.

15. *NASA News Release*, No. 62-257, December 5, 1962.

16. *NASA News Release*, No. 63-186, August 16, 1963.

17. *NASA News*, November 17, 1964, p. 14.

18. In September 1963, President Kennedy suggested a joint moon program and President Johnson reaffirmed this offer. *Department of State Bulletin*, October 7, 1963, pp. 532-33; December 30, 1963, p. 1011.

19. U.S. Congress, Senate, Committee on Aeronautical and Space Sciences, *Hearing, Space Cooperation Between the United States and the Soviet Union*, 92nd Cong., 1st Sess., 1971, pp. 3-7.

20. Ibid., p. 20.

21. United Nations Document A/AC.105/6, July 9, 1962. The Soviet draft also included principles on the rescue of astronauts—an agreed point.

22. Testimony of William Gilbert Carter, House Military Operations Subcommittee, Part 2, p. 665. Much of the following information on U.S. discussions with the USSR comes from this source, pp. 665-66. Also interview, March 18, 1965.

23. Ibid., p. 666

24. Paper read at the International Institute of Space Law, 7 Colloquium on the Law of Outer Space, Warsaw, September 9-11, 1964.

25. See United Nations Document A/AC.105/PV.26-35 for this and the following ideas.

26. *NASA News*, January 25, 1965, pp. 13-14.

27. United Nations Document A/AC.105/46, August 9, 1968.

28. "An Analysis of the Socialist States' Proposal for INTERSPUTNIK: An International Communication Satellite System," *Villanova Law Review*, 15: 1 (Fall, 1969), pp. 83-105, 88.

29. In INTELSAT, this problem is overcome by lowering the majority required for passage on important matters after sixty days. See Agreement, Article V(d).

30. For a detailed analysis of the entire Soviet space program through 1970, see U.S. Congress, Senate, Committee on Aeronautical and Space Sciences, *Soviet Space Programs, 1966-70* 92nd Cong., 1st Sess., 1971.

31. Shillinglaw, "Soviet Union," pp. 221, 94.

32. On November 29, 1965, color television experiments between the USSR and France were initiated using the French color TV system and the Molniya satellite. *Spaceflight*, 8 (January, 1966), p. 1.

33. Just how powerful is a matter of conjecture. Mass media without the use of person-to-person communications may be a poor tool for manipulating attitudes and behavior. For spectrum of opinion, see Ithiel de Sola Pool, "The Mass Media and their Interpersonal Social Functions in the Process of Modernization," in Lewis Anthony Dexter and David M. White, eds., *People, Society, and Mass Communications*, pp. 429-43.

34. In 1968, the General Assembly established a Working Group on Direct Broadcast Satellites as part of the Committee on the Peaceful Uses of Outer Space. Also see U.S. Congress, House of Representatives, Subcommittee on National Security Policy and Scientific Developments, Committee on Foreign Affairs, *Hearings, Satellite Broadcasting: Implications for Foreign Policy*, 91st Cong., 1st Sess., 1969.

35. *Aviation Week and Space Technology*, October 4, 1971, p. 12.

36. *NASA News*, August 1, 1964, p. 12.

37. Ibid.

38. Ibid., p. 13.

39. *The New York Times*, September 23, 1962, p. 64.

40. Marshall Vasili D. Sokolovsky, ed., *Military Strategy, Soviet Doctrine and Concepts* (New York: Praeger, 1963), p. 305.

41. As quoted in *United States Naval Institute Proceedings*, 97: 2/820 (June, 1971), p. 22.

## Chapter 8
## The Transition Between the Interim and
## Definitive Arrangements

1. Senate Commerce Committee, *Incorporation Hearings*, p. 90. Comsat's Presidentially appointed directors are compared to the former Government directors of the Union Pacific Railroad, but this is a bad analogy. For example, see Herman Schwartz, "Governmentally Appointed Directors in a Private Corporation—The Communications Satellite Act of 1962," *Harvard Law Review* 79 (December, 1965), pp. 350-64.

2. Comsat. *Annual Report, 1970*, p. 25, and Communications Satellite Act of 1962 as amended, 76, Stat., 423.

3. This listing comes from "The Company Nobody Knows," *Forbes* (January 15, 1965), pp. 19-20.

4. *Washington Post*, July 15, 1965, p. 313 and John McDonald, "The Comsat Compromise Starts a Revolution," *Fortune* 72 (October, 1965), p. 131.

5. House Military Operations Subcommittee, *Satellite Communications Hearings*, Part 2, p. 742.

6. In 1970, the earth stations constituted approximately 30 percent of the rate base on which the FCC calculated Comsat's profit allowance; but with full ownership, the ground terminals would amount to about 45 percent of the Corporation's rate base. See *Aviation Week & Space Technology*, May 4, 1970, p. 30.

7. Interview, Edwin J. Istvan, March 1, 1966. See also Jack Gould, *New York Times*, April 27, 1965, p. 25.

8. See for instance "Reply Comments of AT&T," FCC, Docket No. 15735, RM-644, January 22, 1965.

9. Katherine Johnsen, "Comsat Leases Rights to Determine Growth of Space Communications," *Aviation Week and Space Technology*, November 8, 1965, pp. 27-28.

10. Quoted in ibid., p. 28.

11. Testimony of the DTM to the House Appropriations Committee, March 7, 1967, p. 7.

12. Mr. Roback has seen the possibility of a conflict of interest here. House Military Operations Subcommittee, *Satellite Communications Hearings*, Part 1, p. 233.

13. U.S. Congress, House, Committee on Government Operations, Military Operations Subcommittee, *Hearings, Government Use of Satellite Communications*, 89th Cong., 2nd Sess., 1966, pp. 393-95, 406-07, hereinafter cited as House Government Operations Committee, *Government Use of Satellite Communications Hearings*.

14. Executive Order 11191, January 4, 1965.

15. Shirley Sheibla, "Comsat Revisited: At Home and Abroad, Its Precise Role is Still Up in the Air," *Barron's* (July 19, 1965), p. 10.

16. The exact procedures by which the FCC and Executive Branch agencies regulate the positions of Comsat within INTELSAT can be found in House Government Operations Committee, *Government Use of Satellite Communications Hearings*, pp. 393-95, 406-07.

17. *The Washington Post*, December 4, 1970, p. A1. The GAO report was requested by Senator Mike Gravel (D., Alaska). It also points out that the authorized user decision is costing the Pentagon at least $3,375,000 a year in extra costs, as the DCA pays commercial carriers $8,250,000 a year for circuits and these carriers in turn pay Comsat only $4,875,000. However, overall rate reductions for all services may justify this. See Asher Ende, "International Telecommunications: Dynamics of Regulation of a Rapidly Expanding Service," *Law and Contemporary Problems*, 34 (Summer, 1969), pp. 389-416, 411.

18. Senate Committee on Aeronautical and Space Sciences, *National Communications Satellite Hearings*, p. 64.

19. Ibid., p. 74.

20. House Military Operations Subcommittee, *Satellite Communications Hearings*, Part 2, p. 744.

21. U.S. Congress, House, *Report on Activities and Accomplishments Under the Communications Satellite Act: Message from the President of the United States*, 89th Cong., 2nd Sess., 1966, H. Doc. 400, p. 2.

22. *House Doc. No. 157*, 90th Cong., 1st Sess., 1967.

23. President's Task Force on Communications Policy, *Final Report* (Washington, D.C., December 7, 1968), hereinafter cited as *Rostow Report*. President Johnson refused to release the report to the public, but in March, 1969, the Nixon Administration reversed this decision.

24. *Rostow Report*, ch. 9 and *New York Times*, March 12, 1969, p. 75M.

25. *Rostow Report*, ch. 2.

26. Ibid., ch. 5.

27. *House Doc. No. 91-222*, 91st Cong., 2nd Sess., 1970. 84 Stat. 2083 (1970). Executive Order 11556, 3 C.F.R. 158 (1970 Comp.).

28. FCC, Docket 16495.

29. *Aviation Week and Space Technology* (November 8, 1971), p. 9.

30. *Broadcasting* (June 14, 1971), pp. 48-50.

31. U.S. Congress, Senate, Committee on Aeronautical and Space Sciences, *International Cooperation in Outer Space: A Symposium*, 92nd Cong., 1st Sess., 1971, pp. 191-94.

32. *Hearings and Report, Assessment of Space Communications Technology*, 91st Cong., 1st Sess., 1969.

33. *Hearings (with Analysis and Findings), Satellite Broadcasting: Implications for Foreign Policy*, 91st Cong., 1st Sess., 1969. *Hearings, Foreign Policy Implications of Satellite Communications*, 91st Cong., 2nd Sess., 1970.

34. *Progress Report on Space Communications*, 89th Cong., 2nd Sess., 1966.

35. Lindblom, *Intelligence of Democracy.*

36. Huntington, *The Common Defense*, pp. 146-47. Cf., ch. I. On the pluralist model of executive-legislative relationships and alternatives, see John S. Saloma, III, *Congress and the New Politics* (Boston: Little, Brown, 1969), ch. 2. Lowi might criticize the policy-making process here as one of "interest group liberalism" which leads to excessive delegation of power without standards. *The End of Liberalism*, p. 144 et passim. Lindblom, on the other hand, would defend the system in terms of incrementalist rationality. In regards to space communications policy of the traditional point-to-point variety, I would favor Lindblom's approach. See Chapter 9.

37. Twentieth Century Fund, *Communicating by Satellite* (New York, 1969), pp. 14-15. Lincoln P. Bloomfield, ed., *Outer Space* (New York: Praeger, 1968), p. 101.

38. *Washington Post* (June 24, 1971), p. E14. *Telecommunication Journal*, 38: 4 (April, 1971), pp. 189-91.

39. To increase the sophistication of the analysis, the relative importance of different structures and functions could be weighted by a figure expressing the average contribution of that structure to the total system and a figure expressing the average level of fulfillment of the function. One would then have a weighted additive index on which one could place different structures and functions, producing a continuum of integration-disintegration. For a fuller exposition of this notion, see Johan Galtung, "A Structural Theory of Integration," *Journal of Peace Research*, No. 4 (1968), pp. 375-95.

40. Ibid., p. 386.

41. TIAS No. 5646 at 5-6. Interview, Stephen E. Doyle (Foreign Affairs Officer, Office of Telecommunications, Department of State), July 23, 1969.

42. TIAS No. 5646 at 77-106.

43. Comsat, *Report to the President and Congress for Calendar Year 1970*, Appendix 6. Also see Appendix B.

44. *Washington Post*, June 24, 1971, p. E14.

45. INTELSAT, *Report of the Interim Communications Satellite Committee on Definitive Arrangements for an International Global Communications Satellite System*, p. 16, hereinafter cited as INTELSAT *Report of Interim Committee on Definitive Arrangements.*

46. The President of the United States, *The Annual Report for 1970 on Activities and Accomplishments under the Communication Satellite Act of 1962*, 92nd Cong., 1st Sess., 1971, p. 12.

47. Comsat, *1970 Annual Report*, p. 46.

48. *Washington Post* (June 24, 1971), p. E14. *Telecommunications Journal*, 38:4 (April, 1971), pp. 189-91, and interview, Stephen Doyle, April 20, 1972.

49. INTELSAT, *Report of Interim Committee on Definitive Arrangements*, p. 23, and Ambassador Abbott Washburn (Chairman U.S. Delegation, INTELSAT Conference), "The International Telecommunications Satellite Con-

sortium," U.S. Congress, Senate, Committee on Aeronautical and Space Sciences, *International Cooperation in Outer Space: A Symposium*, 92nd Cong., 1st Sess., 1971, pp. 437-52, hereinafter cited as Senate Aeronautical and Space Sciences Committee, *International Cooperation in Outerspace Symposium*.

50. Shillinglaw, "Soviet Union."

51. *Business Week*, May 3, 1969, p. 60.

52. Twentieth Century Fund, *Communicating by Satellite* (New York, 1969).

53. However, as a negotiating gambit, the U.S. did not agree to this solution till late 1969, according to Doyle. Cf., Katherine Johnsen, "Comsat, State Department Split of Negotiations," *Aviation Week and Space Technology* (March 31, 1969), pp. 24-25.

54. U.S. Congress, House, Subcommittee of the Committee on Appropriations, *Hearings, Supplemental Appropriation Bill, 1971*, 91st Cong., 2nd Sess., 1970, p. 701.

55. Cf. Herbert I. Schiller, *Mass Communications and American Empire* (Boston: Beacon Press, 1969).

56. Assembly of Western European Union, Committee on Scientific, Technological and Aerospace Questions, *Report, State of European Space Activities–INTELSAT* (Doc. 495), 15th Ordinary Sess., 2nd Part, 1969, pp. 3-6.

57. *ELDO/ESRO Bulletin*, November, 1970, pp. 4-6. Europe may form a European Space Organization (ESO), which could narrow the space technology gap by the late 1970s. See *Aviation Week and Space Technology* (November, 1970), pp. 17-18.

58. Remarks of Senator Clinton Anderson (D-NM), *Congressional Record* (daily edition), November 25, 1969, p. 515027.

59. *Report of the Interim Communications Satellite Committee on Definitive Arrangements for an International Global Communications Satellite System* (Washington, D.C.: INTELSAT, 1968), hereinafter cited as ICSC Report. This listing of issues is my own and reflects my estimation of their order of importance.

60. For a chronological summary, see Washburn, op. cit.

61. France objected to this solution because she wished more power to reside with the Assembly, and thus she abstained in voting for the Definitive Arrangements. See Summary Record, 63rd Plenary Session (SR/63) (Washington, D.C.: INTELSAT, May 21, 1971).

62. *Final Report of the Preparatory Committee (PC(III)/62)*, (Washington, D.C.: INTELSAT, 1969), p. 2.

63. Katherine Johnsen, "Effort to Cut Comsat's Powers Foreseen," *Aviation Week & Space Technology*, February 24, 1969, pp. 22-23.

64. *Report of the United States Delegation to the Plenipotentiary Conference on Definitive Arrangements for the International Telecommunications Satellite Consortium* (First Session), (Washington, D.C., April 10, 1969).

65. Ibid.

66. Resumed Plenipotentiary Conference on Definitive Arrangements for the International Telecommunications Satellite Consortium, "An Outline of a Proposed Resolution of Certain Basic Issues" (Doc. 93) (Washington, D.C.: INTELSAT, March 8, 1970). Other aspects of the package included voting in the Board of Governors, the basis of investment, and the structure of the organization.

67. Katherine Johnsen, "U.S. Accedes to Establishment of Regional Satellite Networks," *Aviation Week & Space Technology* (December 22, 1969), p. 23 and Final Report of the Preparatory Committee, passim.

68. As quoted in Katherine Johnsen, "U.S., U.K. Fight over Services Stalls Permanent Intelsat Accords," *Aviation Week & Space Technology* (May 17, 1971), p. 19.

69. Ibid.

70. *Business Week*, June 14, 1969, p. 50.

71. *ICSC Report*, p. 81.

72. *Washington Post*, June 23, 1969, p. D11. The pressure to share contracts according to some egalitarian formula has undermined ELDO and EURATOM.

73. See *Aviation Week & Space Technology* (October 12, 1970), pp. 21-22 and January 18, 1971, p. 13.

74. Resumed Plenipotentiary Conference on Definitive Arrangements for the International Telecommunications Satellite Consortium, "Remarks of the President of the United States," (Washington, D.C.: INTELSAT, May 21, 1971).

75. EUROSPACE, *Memorandum, The Urgent Measures Required for the Implementation of a European Space Programme* (Paris, 1966), pp. 6-7.

76. *Business Week*, December 25, 1965, p. 17.

77. House Subcommittee on Space Science and Applications, *Assessment of Space Communications Technology Hearings*, pp. 12-13.

78. House Subcommittee on National Security Policy and Scientific Developments, *Foreign Policy Implications of Satellite Communications Hearings*, p. 1.

79. House Subcommittee on National Security Policy and Scientific Developments, *Reports of the Special Study Mission to Latin America*, 91st Cong., 2nd Sess., 1970, pp. 35-38.

80. "Challenge to Cooperation," *Saturday Review* (October 24, 1970), p. 25.

81. *Telecommunications Journal*, 38: 1 (October, 1971), pp. 673-82.

82. Paul F. Geven, "Worldwide Standards for Color Television," Department of State Bulletin, 53: 1372 (October 11, 1965).

83. Comsat, *Report to the President and Congress* (Washington, D.C., April 20, 1970), pp. 19-28.

84. United Nations, Committee on the Peaceful Uses of Outer Space, *Report of the Working Group on Direct Broadcast Satellites on its Third Session*, Doc. A/AC-105/83, May 25, 1970.

85. Library of Congress, Legislative Reference Service, *Direct Broadcasting from Satellites: Annotated Bibliography* (Washington, D.C., July 15, 1968).

86. 'National Citizens' Commission, *Committee on Communications*, pp. 20-23.

87. Committee on the Peaceful Uses of Outer Space, Working Group on Direct Broadcast Satellites, p. 7.

88. Lasswell, "Science, Scientists and World Policy."

89. See Isaac Asimov, "The Fourth Revolution," *Saturday Review* (October 24, 1970), pp. 17-20. But mass communications alone will not change attitudes. They may reinforce existing prejudices. See Leigh, "International Relations."

**Chapter 9**
**Conclusions**

1. Huntington, *The Common Defense*, pp. 146-47.

2. "The Economics of Knowledge and the Knowledge of Economics," *American Economic Review*, 56 (May, 1966), pp. 1-13, 11.

3. Kenneth N. Waltz, *Man, the State and War* (New York: Columbia University Press, 1959).

4. Lindblom, *The Intelligence of Democracy*, p. 151 et passim.

5. Ibid.

6. James R. Schlesinger, "Quantitative Analysis and National Security," *World Politics*, 15 (January, 1963), pp. 295-315.

7. E.g., Zbigniew Brzezinski and Samuel P. Huntington, *Political Power: USA/USSR* (New York: The Viking Press, 1964), ch. 4 and James A. Robinson, *Congress and Foreign Policy-Making* (Homewood, Ill.: The Dorsey Press, 1967).

8. Ronald C. Moe and Steven C. Teel draw our attention to the neglected initiatory role of Congress. "Congress as Policy-Maker: A Necessary Reappraisal," *Political Science Quarterly*, 85 (September, 1970), pp. 443-70.

9. Rosenau, "Pre-theories and Theories of Foreign Policy."

# Glossary

| | |
|---|---|
| ABC | American Broadcasting Company |
| ADVENT | A still-born Defense Communication Satellite System |
| ARPA | Advanced Research Projects Agency |
| ATS | Applications Technology Satellites |
| AT&T | American Telephone and Telegraph Company |
| CBS | Columbia Broadcasting System |
| CEPT | European Conference of Postal and Telecommunications Administrations |
| CETS | Conference Europeenne des Telecommunications par Satellites (Also ECSC) |
| COMSAT | Communications Satellite Corporation |
| COURIER | An Army active communications satellite |
| DCA | Defense Communications Agency |
| DSCS | Defense Satellite Communications System |
| DTM | Director of Telecommunications Management |
| EARC | Extraordinary Administrative Radio Conference—1963 |
| EARLY BIRD | INTELSAT I |
| ECHO | A passive communication system employing balloons |
| ECSC | European Conference on Satellite Communications (Also CETS) |
| ELDO | European Space Vehicle Launcher Development Organization |

| | |
|---|---|
| ESRO | European Space Research Organization |
| EUROSAT | A consortium of European firms which plans to function as the manager and operator of a European-regional satellite system |
| EUROSPACE | European Industrial Space Study Group |
| FCC | Federal Communications Commission |
| FLEETSAT | Fleet Satellite Communications System |
| GE | General Electric |
| GT&E | General Telephone and Electric Company |
| HTC | Hawaiian Telephone Company |
| IDCSP | Interim Defense Communications Satellite Project |
| IFRB | International Frequency Registration Board |
| IGY | International Geophysical Year |
| INTELSAT | International Telecommunications Satellite Consortium (Since 1972 Organization rather than Consortium) |
| INTELSAT I,II,III,IV | The names of the four generations of Comsat satellites |
| INTERSPUTNIK | The proposed Communist International Communications Satellite System |
| IRAC | Interdepartment Radio Advisory Committee |
| ITT | International Telephone and Telegraph |
| ITT Worldcom | ITT World Communications, Inc. |
| ITU | International Telecommunications Union |
| MACS | Medium-Altitude Communications System |

| | |
|---|---|
| MOLNIYA | "Lightning": The first generation Russian communications satellites |
| NACA | National Advisory Committee for Aeronautics |
| NASA | National Aeronautics and Space Administration |
| NASC | National Aeronautics and Space Council |
| NASCOM | Satellite communications program for moon landing |
| NBC | National Broadcasting Company |
| NCS | National Communications System |
| OCDM | Office of Civil and Defense Mobilization |
| OEP | Office of Emergency Planning |
| ORBITA | The Russian Domestic Communications Satellite System |
| OTP | Office of Telecommunications Policy |
| RCA | Radio Corporation of America |
| RCAGC | RCA Global Communications |
| RELAY | An early NASA active experimental communications satellite |
| SCORE | The first active communications satellite (1958– Army) |
| SKYNET | British-American military communication satellite program |
| STATSIONAR | The second generation Russian communication satellites |
| SYMPHONIE | A French-German experimental communications satellite program |

| | |
|---|---|
| SYNCOM | The first synchronous satellite |
| TELSTAR | The Bell System's 1962 active communications satellite |
| USIA | United States Information Agency |
| WARC | World Administrative Radio Conference—1971 |
| West Ford | An experimental defense passive communications system |
| WU | Western Union |
| WUI | Western Union International |

# Bibliography

**Public Documents**

Assembly of Western European Union, Committee on Scientific, Technological and Aerospace Questions, *Report on State of European Space Activities– INTELSAT* (Document 495). 15th Ordinary Session, 2nd Part, 1969.

Gardner, Richard N. "Space Meteorology and Communications: A Challenge to Science and Diplomacy," *Department of State Bulletin*, 48, No. 1246 (May 13, 1963).

Geren, Paul F. "Worldwide Standards for Color Television," *Department of State Bulletin*, 53, No. 1372 (October 11, 1965).

INTELSAT. *Report of the ICSC on Definitive Arrangements for the International Telecommunications Satellite Consortium*. Washington, December 31, 1968.

_____. *Final Report of the Preparatory Committee of the Plenipotentiary Conference on Definitive Arrangements for the International Telecommunications Satellite Consortium*. Washington, December 11, 1969.

_____. *Summary Records, Working Documents*, and *Committee Reports of the Plenipotentiary Conference on Definitive Arrangements for the International Telecommunications Satellite Consortium*. Washington, 1969-1971.

_____. *Reports of the U.S. Delegation to the Plenipotentiary Conference on Definitive Arrangements for the International Telecommunications Satellite Consortium*. Washington, 1969-1971.

International Telecommunications Union, *Telecommunications in the Space Age*. Paris, 1966.

Loevinger, Lee. "Cooperation in International Communications," *Department of State Bulletin*, 53, No. 1378 (November 22, 1965).

The President's Communications Policy Board. *Telecommunications–A Program for Progress*. Washington: Government Printing Office, 1951.

United Kingdom, Central Office of Information. *Britain and Commonwealth Telecommunications*. London: British Information Services, 1963.

United Kingdom, Central Office of Information. *Britain and Space Research*. London: British Information Services, 1965.

United Nations. Doc. No. A/AC. 105/OR.2. New York: United Nations, March 29, 1963.

United Nations. Doc. No. A/AC.105/PV4. New York: United Nations, March 21, 1962.

United Nations. General Assembly. Committee on the Peaceful Uses of Outer Space, *Report of the Working Group on Direct Broadcast Satellites on Its Third Session*. (Document No. A/AC. 105/83). May 25, 1970.

U.S. Department of State. *Report of the United States Delegation to the Plenipotentiary Conference of the ITU. 1965*. Washington, 1966.

U.S. Federal Communications Commission. *FCC Annual Reports for each Fiscal Year*. Washington, D.C.: Government Printing Office.

U.S. General Accounting Office. *Report to the Congress on Large Costs to Government Not Recovered for Launch Services Provided to the Communications Satellite Corporation* (B-168707). October 8, 1971.

U.S. House of Representatives. Committee on Foreign Affairs, Subcommittee on International Organizations and Movements. *Modern Communications and Foreign Policy*. 90th Cong., 1st Sess., 1967.

U.S. House of Representatives. Committee on Foreign Affairs, Subcommittee on National Security Policy and Scientific Developments. *Hearings on Satellite Broadcasting: Implications for Foreign Policy*. 91st Cong., 1st Sess., 1969.

U.S. House of Representatives. Committee on Foreign Affairs, Subcommittee on National Security Policy and Scientific Developments. *Hearings on Foreign Policy Implications of Satellite Communications*. 91st Cong., 2nd Sess., 1970.

U.S. House of Representatives. Committee on Government Operations. *Government Operations in Space*. Report No. 445. 89th Cong., 1st Sess., 1965.

U.S. House of Representatives. Committee on Government Operations, Military Operations Subcommittee. *Hearings on Missile and Space Ground Support Operations*. 89th Cong., 2nd Sess., 1966.

U.S. House of Representatives. Committee on Government Operations, Military Operations Subcommittee. *Hearings on Satellite Communications–1964*. Parts 1 and 2, 88th Cong., 2nd Sess. 1964.

U.S. House of Representatives. Committee on Government Operations. *Missile and Space Ground Support Operations*. Report No. 1340. 89th Cong., 2nd Sess., 1966.

U.S. House of Representatives. Committee on Government Operations, Military Operations Subcommittee. *Satellite Communications*. Report. 88th Cong., 2nd Sess., 1964.

U.S. House of Representatives. Committee on Government Operations, Military Operations Subcommittee. *Hearings on Government Use of Satellite Communications*. 89th Cong., 2nd Sess., 1966.

U.S. House of Representatives. Committee on Government Operations. *Forty-Third Report, Government Use of Satellite Communications*. 89th Cong., 2nd Sess., 1966.

U.S. House of Representatives. Committee on Government Operations, Military Operations Subcommittee. *Hearings on Government Use of Satellite Communications–1967*. 90th Cong., 1st Sess., 1967.

U.S. House of Representatives. Committee on Government Operations. *Seventh Report, Government Use of Satellite Communications–1967*. 90th Cong., 1st Sess., 1967.

U.S. House of Representatives. Committee on Government Operations, Military Operations Subcommittee. *Hearings on Military Communications–1968*. 90th Cong., 2nd Sess., 1968.

U.S. House of Representatives. Committee on Government Operations. *Thirty-Fourth Report, Government Use of Satellite Communications—1968*. 90th Cong., 2nd Sess., 1968.

U.S. House of Representatives. Committee on Interstate and Foreign Commerce. *Communications Satellite Act of 1962—The First Year*. Report No. 809. 88th Cong., 1st Sess., 1963.

U.S. House of Representatives. Committee on Interstate and Foreign Commerce. *Hearings on Communications Satellites*. Parts 1 and 2, 87th Cong., 1st Sess., 1961.

U.S. House of Representatives. Committee on Science and Astronautics. *Commercial Applications of Space Communications Systems*. Report No. 1279. 87th Cong., 1st Sess., 1961.

U.S. House of Representatives. Committee on Interstate and Foreign Commerce. *Communications Satellite Act of 1962*. Report No. 1636 to accompany H.R. 11040. 87th Cong., 2nd Sess., 1962.

U.S. House of Representatives. Committee on Science and Astronautics. *Hearings on Communications Satellites*. Parts 1 and 2, 87th Cong., 1st Sess., 1961.

U.S. House of Representatives. Committee on Science and Astronautics. *Hearings on Satellites for World Communications*. 86th Cong., 1st Sess., 1959.

U.S. House of Representatives. Committee on Science and Astronautics. *Proposed Studies on the Implications of Peaceful Space Activities for Human Affairs*. Report No. 242. 87th Cong., 1st Sess., 1961.

U.S. House of Representatives. Committee on Science and Astronautics, Subcommittee on Space Science and Applications. *Hearings on Assessment of Space Communications Technology*. 91st Cong., 1st Sess., 1969.

U.S. House of Representatives. Committee on Science and Astronautics, Subcommittee on Space Science and Applications, *Report on Assessment of Space Communications Technology*. 91st Cong., 1st Sess., 1970.

U.S. House of Representatives. Committee on Science and Astronautics, Panel on Science and Technology, Thirteenth Meeting. *Space Communications*, (a paper by Carl Hammer). 92nd Cong., 2nd Sess., 1972.

U.S. National Aeronautics and Space Administration. *Conference on the Law of Space and of Satellite Communications*. Doc. No. SP-84, Washington: Government Printing Office, 1964.

U.S. National Aeronautics and Space Administration. *NASA News Release*, No. 63-186, August 16, 1963.

U.S. *Papers Prepared for the U.N. Conference on the Application of Science and Technology for the Benefit of the Less Developed Area*. Vol. 12: *Communications*. Washington: Government Printing Office, 1963.

U.S. Post Office Department. *Government Ownership of Electrical Means of Communication*. Doc. No. 399. 63rd Cong., 2nd Sess., 1914.

U.S. President. *Reports on Activities and Accomplishments Under the Communications Satellite Act*. Printed as a House Document in each Congress from the 88th Cong., 1st Sess., 1963 to present.

U.S. President. (Johnson). *Global Communication System*. House Document No. 157. 90th Cong., 1st Sess., 1967.

U.S. President. (Johnson). "Report of the Task Force on Communications Policy " December 7, 1968.

U.S. Senate, Committee on Aeronautical and Space Sciences. *Communications Satellite Act of 1962*. Report No. 1319. 87th Cong., 2nd Sess., 1962.

U.S. Senate Committee on Aeronautical and Space Sciences. *Communications Satellites: Technical, Economic, and International Developments*. Staff Report. 87th Cong., 2nd Sess., 1962.

U.S. Senate, Committee on Aeronautical and Space Sciences. *Hearings on Communications Satellite Legislation*. 87th Cong., 2nd Sess., 1962.

U.S. Senate Committee on Aeronautical and Space Sciences. *Hearings on National Communications Satellite Programs*. 89th Cong., 2nd Sess., 1966.

U.S. Senate, Committee on Aeronautical and Space Sciences. *Hearing on Nomination of Incorporators*. 88th Cong., 2nd Sess., 1963.

U.S. Senate, Committee on Aeronautical and Space Sciences. *International Cooperation and Organization for Outer Space*. Report No. 56. 89th Cong., 1st Sess., 1965.

U.S. Senate, Committee on Aeronautical and Space Sciences. *Policy Planning for Space Telecommunications*. Staff Report. 86th Cong., 2nd Sess., 1960.

U.S. Senate, Committee on Aeronautical and Space Sciences. *Radio Frequency Control in Space Telecommunications* Committee Print, 86th Cong., 2nd Sess., 1960.

U.S. Senate, Committee on Aeronautical and Space Sciences. *Soviet Space Programs: Organization, Plans, Goals, and International Implications*. Staff Report. 87th Cong., 2nd Sess., 1962.

U.S. Senate, Committee on Aeronautical and Space Sciences. *Staff Report, Soviet Space Programs, 1966-70*. 92nd Cong., 1st Sess., 1971.

U.S. Senate. Committee on Aeronautical and Space Sciences. *International Cooperation in Outer Space: A Symposium*. 92nd Cong., 1st Sess., 1971.

U.S. Senate, Committee on Commerce. *Communications Satellite Act of 1962*. Report No. 1584 to accompany H.R. 11040. 87th Cong., 2nd Sess., 1962.

U.S. Senate Committee on Commerce. *Hearings on Communications Satellite Incorporators*. 88th Cong., 1st Sess., 1963.

U.S. Senate, Committee on Commerce. *Hearings of Communications Satellite Legislation*. 87th Cong., 2nd Sess., 1962.

U.S. Senate, Committee on Commerce, Communications Subcommittee. *Hearings on Satellite Communications*. 88th Cong., 1st Sess., 1963.

U.S. Senate, Committee on Commerce, Communications Subcommittee. *Hearings on Space Communications and Allocation of Radio Spectrum*. 87th Cong., 1st Sess., 1961.

U.S. Senate. Committee on Commerce, Subcommittee on Communications. *Hearings on Progress Report on Space Communications*. 89th Cong., 2nd Sess., 1966.

U.S. Senate, Committee on Foreign Relations. *Communications Satellite Act of 1962*. Report No. 1873 to accompany H.R. 11040. 87th Cong., 2nd Sess., 1962.

U.S. Senate, Committee on Foreign Relations. *Hearings on Communications Satellite Act of 1962*. 87th Cong., 2nd Sess., 1962.

U.S. Senate, Select Committee on Small Business, Subcommittee on Monopoly. *Hearings on Space Satellite Communications* Parts 1 and 2, 87th Cong., 1st Sess., 1961.

U.S. Senate, Committee on Interstate Commerce, A Subcommittee. *Hearings on Cable Landing Licenses*. 66th Cong., 3rd Sess., 1921.

U.S. Senate, Committee on Interstate Commerce, Subcommittee. *Hearings on Study of International Communications*. Parts 1 and 2, 79th Cong., 1st Sess., 1945.

U.S. Senate, Committee on Interstate Commerce. *Investigation of International Communications by Wire and Radio*. Report No. 1907. 79th Cong., 2nd Sess., 1946.

U.S. Senate, Committee on Interstate Commerce. *Investigations of International Communications by Wire and Radio*. Report No. 19, 80th Cong., 1st Sess., 1947.

U.S. Senate, Committee on the Judiciary, Antitrust and Monopoly Subcommittee. *Hearings on Antitrust Problems of the Space Satellite Communications System*. Parts 1 and 2, 87th Cong., 2nd Sess., 1962.

**Interviews**

Aviation Week and Space Technology. Katherine Johnsen, Reporter, Washington, D.C. April 22, 1966.

Communications Satellite Corporation. Richard Colino, International Developments, Washington, D.C. April 22, 1965.

_____. Edwin J. Istvan, Director of International Developments, Washington, D.C. March 1, 1966 and July 8, 1966.

_____. John A. Johnson, Vice President, International, Washington, D.C. April 22, 1965.

_____. David Leive, Lawyer, Washington, D.C. April 22, 1965 and March 1 and June 8, 1966.

_____. F. John D. Taylor, Special Assistant to Vice President, Planning, Washington, D.C. March 21, 1966.

Italy. Dr. Franco Fiorio, Vice Chairman, Subcommittee for Contracts, INTELSAT, Arlington, Virginia, May 11, 1966.

U.S. Congress, Senate, Committee on Aeronautical and Space Sciences. Eilene Galloway, Special Consultant, Washington, D.C., December, 1965.

U.S. Congress, Senate, Committee on Commerce. Nicholas Zapple, Communications Council, Washington, D.C., November 8, 1965.

U.S. Congress, House, Committee on Government Operations, Military Operations Subcommittee. Daniel W. Fulmer, Attorney, Washington, D.C. March 16, 1965, February 23 and March 11, 1966.

——. Herbert Roback. Staff Director. Washington, D.C. September 6, 1968.

U.S. Congress, House, Committee on Interstate and Foreign Commerce. Kurt Borchardt, Staff, Washington, D.C. February 22 and March 6, 1966.

U.S. Department of State. William Gilbert Carter, Special Assistant for Space Communications, Bureau of Economic Affairs 1962-64, Washington, D.C. March 18, 1965 and July 28, 1966.

——. Francis Colt de Wolf, Chief Telecommunications Division, 1935-1964 (retired), Washington, D.C. April 22, 1965.

——. Thomas E. Nelson, Telecommunications Division, Bureau of Economic Affairs, Washington, D.C. April 20, 1966.

——. T.H.E. Nesbitt, Deputy Chief, Outer Space Affairs, Office of International Scientific and Technological Affairs, Washington, D.C. April 21, 1965.

——. Robert Packard, Chief, Outer Space Affairs, Office of International Scientific and Technological Affairs, Washington, D.C. April 19, 1965.

——. Stephen E. Doyle, Foreign Affairs Officer, Officer of Telecommunications. Washington, D.C. July 23, 1969 and August 21, 1970.

U.S. Federal Communications Commission. Lee Loevinger, Commissioner, Washington, D.C. October 26, 1965.

——. Asher Ende, Deputy Chief, Common Carrier Bureau. Washington, D.C. May 2, 1966 and January 15, 1971.

U.S. National Aeronautics and Space Council. M.V. Mrozinski, Staff, Washington, D.C. April 16, 1965.

——. Charles S. Sheldon, II. Staff, Washington, D.C. March 17, 1965.

——. Edward C. Welsh, Executive Secretary, Washington, D.C. October 12, 1965.

U.S. Office of Emergency Planning. Ralph L. Clark, Chief, Telecommunications Management Division, Washington, D.C. October 4, 1965, and July 26, 1966.

U.S. Office of Telecommunications Policy. Stephen E. Doyle, Staff. Washington, D.C., April 20, 1972.

——. Charles C. Joyce, Assistant to the Director, Washington, D.C. April 21, 1972.

## Reports

Carnegie Endowment for International Peace and the Twentieth Century Fund. *Communicating by Satellite: An International Discussion*. New York, 1969.

——. *Planning for a Planet: An International Discussion on the Structure of Satellite Communications*. New York, 1971.

Communications Satellite Corporation. *Reports Pursuant to Section 404(b) of the Communications Satellite Act of 1962; 1963-1971*. Washington, D.C., 1964-1972.

Communications Satellite Corporation. European Office. *European Reports*. Geneva, November 11, 1968 and January 1, 1970.

EUROSPACE. *Recommendations Concerning the Setting Up of a European Regional Telecommunication Satellite System*. Paris, October, 1967.

National Citizens' Commission. *Report of the Committee on Communications*. The White House Conference on International Cooperation. Washington, D.C., 1965.

_____. *Report of the Committee on Science and Technology*. The White House Conference on International Cooperation. Washington, D.C., 1965.

Schwartz, Murray L., and Goldsen, Joseph M. *Foreign Participation in Communications Satellite Systems: Implications of the Communications Satellite Act of 1962*. Doc. No. RM-3484. Santa Monica, Calif.: The RAND Corp., February, 1963.

Stanford Research Institute. *Possible Nonmilitary Scientific Developments and their Potential Impact on Foreign Policy Problems of the United States*. A Report published as Study No. 2 of the Senate Foreign Relations Committee, *United States Foreign Policy*, 86th Cong., 1st Sess., 1959.

Stockholm International Peace Research Institute. *Communication Satellites*. Stockholm, 1969.

Twentieth Century Fund. *Communicating by Satellite*. New York, 1969.

_____. *The Future of Satellite Communications*. New York, 1970.

UNESCO. *Space Communications and the Mass Media*. Paris, 1963.

**Books**

Bloomfield, Lincoln P. (ed.). *Outer Space*. New York: Praeger, 1968.

Clark, Keith. *International Communications: The American Attitude*. New York: Columbia University Press, 1931.

Creel, George. *How We Advertise America*. New York: Harper, 1920.

Dexter, Lewis A., and White, David M. (eds.). *People, Society, and Mass Communications*. New York: Free Press, 1964.

Dunn, Frederick Sherwood. *War and the Minds of Men*. New York: Harper, 1950.

*From Semaphore to Satellite*. International Telecommunication Union. Geneva: International Telecommunication Union, 1965.

Goldsen, Joseph M. (ed.) *Outer Space in World Politics*. New York: Praeger, 1963.

Haley, Andrew G. *Space Law and Government*. New York: Appleton-Century Crofts, 1963.

Lasswell, Harold D. *Propaganda Technique in the World War*. London: Kegan Paul, 1938.

Leive, David M. *International Telecommunications and International Law: The Regulation of the Radio Spectrum*. Dobbs Ferry, N.Y.: Oceana Publications, 1971.

Lerner, Daniel (ed.). *Propaganda in War and Crisis*. New York: George W. Stewart, 1951.

Lindblom, Charles E. *The Intelligence of Democracy: Decision Making through Mutual Adjustment*. New York: The Free Press, 1965.

Lowi, Theodore J. *The End of Liberalism*. New York: W.W. Norton, 1969.

Mazlish, Bruce. (ed.). *The Railroad and the Space Program: An Exploration in Historical Analogy*. Cambridge: M.I.T. Press, 1965.

McLuhan, Marshall. *Understanding Media: The Extensions of Man*. New York: McGraw-Hill, 1965.

Russett, Bruce M., et al. *World Handbook of Political and Social Indicators*. New Haven: Yale University Press, 1964.

Saloma, John S., III. *Congress and the New Politics*. Boston: Little, Brown, 1969.

Schiller, Herbert I. *Mass Communications and American Empire*. Boston: Beacon Press, 1969.

Schramm, Wilbur L. *Mass Media and National Development: The Role of Information in the Developing Countries*. Stanford: Stanford University Press, 1964.

Sokolovsky, Marshall Vasili D. (ed.). *Military Strategy, Soviet Doctrine and Concepts*. New York: Praeger, 1963.

Taubenfeld, Howard J. (ed.). *Space and Society*. Dobbs Ferry, N.Y.: Oceana Publications, 1964.

Van Dyke, Vernon. *Pride and Power: The Rationale of the Space Program*. Urbana: University of Illinois Press, 1964.

Whitton, John B., and Larson, Arthur. *Propaganda: Towards Disarmament in the War of Words*. Dobbs Ferry, New York: Oceana Publications, 1963.

Wright, Charles R. *Alternative Systems of Mass Communications*. New York: Random House 1959.

## Articles, Periodicals and Newspapers

Asimov, Isaac. "The Fourth Revolution," *Saturday Review*, October 24, 1970, 17-20.

*Aviation Week & Space Technology*. 1962-1972.

*Business Week*. 1964-1971.

Chayes, Abram and Chazen, Leonard. "Policy Problems in Direct Broadcast from Satellites, *Stanford Journal of International Studies*, 5 (June, 1970), 4-20.

Clarke, Arthur C. "Extraterrestrial Relays," *Wireless World*, 51, No.10 (October, 1945), 305-308.

Colino, Richard R. "INTELSAT: Doing Business in Outer Space," *The Columbia Journal of Transnational Law*, 6, No. 1 (Spring, 1967), 17-60.

_____. "International Satellite Telecommunications and Developing Countries," *The Journal of Law and Economic Development*, 3, No. 1 (Spring, 1968), 8-41.

"Corporations Formed Pursuant to Treaty," Note, *Harvard Law Review*, 76, No. 7 (May, 1963), 1431-1449.

Delorme, Jean. "European Space Policy," *Space Digest*, 8, No. 6 (June 1955), 59-61.

Deutsch, Karl W. "The Impact of Science and Technology on International Politics," *Daedalus*, 88, No. 4 (Fall, 1959), 669-685.

*ESRO/ELDO Bulletin*. 1970.

Doyle, Stephen E. "An Analysis of the Socialist States' Proposal for INTER-SPUTNIK: An International Communications Satellite System," *Villanova Law Review*, 15, No. 1 (Fall, 1969), 83-105.

_____. "Communications in the United States of America: Government and Industry Structure," *ITU Telecommunication Journal*, (June, 1968), 1-6.

Ende, Asher. "International Telecommunications: Dynamics of Regulation of a Rapidly Expanding Service," *Law and Contemporary Problems*, 34 (Summer, 1969), 389-416.

Fawcett, J.E.S. "Satellite Broadcasting," *World Today*, 27 (February, 1971). 76-81.

Galloway, Jonathan F. "Worldwide Corporations and International Integration: The Case of INTELSAT," *International Organization*, 24, No. 3 (Summer, 1970), 503-519.

_____. "Technological Change and the Prospects for Democratic Politics" *Journal of Social Philosophy*, 3, No. 2 (April, 1972), 12-15.

Glazer, J. Henry. "Infelix ITU—The Need for Space Age Revisions to the International Telecommunications Conventions," Reprint from the *Federal Bar Journal*, 23, No. 1 (Winter, 1963), 1-36.

Gould, Jack. "Comsat Is Facing Equal-Time Issue," *New York Times* (New York), May 8, 1965.

Grandin, Thomas. "The Political Uses of the Radio," *Geneva Studies*, 10, No. 3 (Geneva, 1939).

Hall, R. Cargill. "Origins and Development of the Vanguard and Explorer Satellite Programs," *The Air Power Historian*, 11, No. 4 (October, 1964), 101-112.

Hurley, Neil P., S.J. "Satellite Communications: A Case Study of Technology's Impact on Politics," *The Review of Politics*, 30, No. 2 (April, 1968), 170-190.

Johnsen, Katherine. "Comsat Leases Rights to Determine Growth of Space Communications," *Aviation Week and Space Technology*, 83, No. 19 (November 8, 1965), 27-28.

———. "Comsat Pushes New Domestic Satellite Net," *Aviation Week and Space Technology*, 83, No. 6 (April 18, 1966), 27-28.

Krieger, Susan. "Prospects for Communication Policy," *Policy Sciences*, 2, No. 3 (Summer, 1971), 305-319.

La Fond, Charles D. "Enthusiasm Grows for Global Comsats," *Missiles and Rockets*, 18, No. 5 (January 31, 1966), 46-55.

Lasswell, Harold D. "Science, Scientists and World Policy," in *Science and the Future of Mankind*, edited by Hugh Boyko. Bloomington: Indiana University Press, 1961.

Leigh, Robert D. "The Mass-Communications Inventions and International Relations," in *Technology and International Relations*, edited by William F. Ogburn. Chicago: Chicago University Press, 1949.

Lerner, Daniel. "Communication Systems and Social Systems," in *Mass Communications*, edited by Wilbur L. Schramm. 2nd edition. Urbana: University of Illinois Press, 1960.

Lowe, Erhard. "Proposals for European Space Projects," *Space Digest*, 8, No. 6 (June, 1965), 62-66.

Lindblom, Charles E. "The Science of 'Muddling Through'," *Public Administration Review*, 29, No. 2 (Spring 1959), 79-85.

Maddox, Brenda. "The Politics of INTELSAT," *The Nation*, February 28, 1972, 272-276.

McDonald, John. "The Comsat Compromise Starts a Revolution," *Fortune*, 72, No. 4 (October, 1965), 131.

McElheny, Victor K. "Is French Scientific Policy Chauvinist?," *Science*, 149, No. 3689 (September 10, 1965), 1217.

Meckling, William. "Economic Potential of Communications Satellites," *Science*, 133, No. 3468 (June 16, 1961), 1885-1892.

Miller, Stewart E. "Communication by Laser," *Scientific American*, 214, No. 1 (January, 1966), 19-27.

*Missiles and Rockets*. Vol. 18, No. 5 (January 31, 1966).

Moe, Ronald C. and Teel, Steven C. "Congress as Policy-Maker: A Necessary Reappraisal," *Political Science Quarterly*, 85, No. 3 (September, 1970), 443-470.

Murphy, Thomas P. "Technology and Political Change: The Public Interest Impact of Comsat," *The Review of Politics*, 33, No. 3 (July, 1971), 405-424.

*New York Times*. 1963-1972.

Riegel, O.W. "Communications by Satellite," *The Quarterly Review of Economics and Business*, 11, No. 4 (Winter, 1971), 23-36.

Rogers, Walter S. "International Electrical Communications," Foreign Affairs, 1, No. 2 (December, 1922), 144-157.

Schramm, Wilbur. "Social and Educational Implications of Communications Satellites," *School & Society* (October 29, 1966), 346-348.

Schwartz, Herman. "Comsat, the Carriers, and the Earth Stations: Some

Problems with 'Melding Variegated Interests,' " *The Yale Law Review*, 76 (January, 1967), 441-484.

Sheibla, Shirley. "Comsat Revisited: At Home and Abroad, Its Precise Role Is Still Up in the Air," *Barron's*, 45 (July 19, 1965), 10.

Shepard, Leslie R. "A Space Programme for Europe," *Spaceflight*, 8, No. 1 (January, 1966), 2.

Shillinglaw, Thomas L. "The Soviet Union and International Satellite Telecommunications," *Stanford Journal of International Studies*, 5 (June, 1970), 199-226.

Smith, Delbert D. "Educational Satellite Telecommunication: The Challenge of a New Technology," *Bulletin of Atomic Scientists*, 27 (April, 1971), 14-18.

Smythe, Dallas W. "Communications Satellites," *Bulletin of the Atomic Scientists*, 17, No. 2 (February, 1961), 67.

Stern, Philip M. "Issues Not Faced in the Telestar Filibuster," *The Reporter*, 27, No. 4 (September 13, 1962), 36-38.

*Telecommunications Reports*. 1968-1971.

"The Peers Debate Space Technology," *Spaceflight*, 8, No. 3 (March, 1966), 102-103.

Throop, Allen. "Some Legal Facets of Satellite Communication," *American University Law Review*, 17, (December, 1967), 12-40.

Trooboff, Peter D. "INTELSAT: Approaches to the Renegotiation," *Harvard International Law Journal*, 9, No. 1 (Winter, 1968), 1-84.

von Neumann, John. "Can We Survive Technology?," *Fortune*, 51, No. 6 (June, 1955), 152.

*Washington Post*. 1964-1972.

Wetmore, Warren C. "$460 Million German Space Push Urged," *Aviation Week and Space Technology*, 83, No. 10 (September 6, 1965), 50-54.

## Miscellaneous Sources

American Institute of Aeronautics and Astronautics. Communication Satellite Systems Conference, Panels on Political, Social, Economic and Legal Problems, Washington, D.C. May 2-4, 1966.

_____. 2nd Communication Satellite Systems Conference. Papers from Panels on Political, Social, Economic, and Legal Problems. San Francisco. April 8-10, 1968.

Doyle, Stephen E. "Permanent Arrangements for the Global Commercial Communication Satellite System of INTELSAT." A paper presented at the 14th Colloquium of the Law of Outer Space During the 22nd International Astronautical Federation. Brussels. September 25, 1971.

Finch, Kenneth Anderson. "Space Communications: Catalyst for International Understanding." A paper presented to the 6th Colloquium on the Law of Outer Space. Paris, 1963, during the 14th Congress of the International Astronautical Federation. (Mimeographed.)

# Index

ADVENT, 20, 32, 36, 62, 105–6, 109, 119
AFL-CIO, 62
AID, 61
AT&T, 11, 12, 15–16, 18, 23, 25, 28, 29, 30, 33, 36, 37, 39, 41, 47, 51, 54, 62, 67, 93, 99, 138, 139, 141, 143, 149, 178
Ad Hoc Carrier Committee, 39, 40, 41, 42, 51
Ad Hoc Committee on the Peaceful Uses of Outer Space, 18
Ad Hoc Communications Satellite Group, 83, 84
Aerospace Corporation, 109–10
Administrative Radio Conference, 75
Advanced Research Projects Agency (ARPA), 13, 20, 108
Agreement on Assistance to and Recovery of Astronauts, 125–26
Agreement Relating to the International Telecommunications Satellite Organization, 158
Agreement on the Rescue of Astronauts and the Return of Objects Launched into Outer Space, 124, 126
Air Force, 20, 21
Alexanderson alternator, 90
"All Red" cable, 9
American Broadcasting Company (ABC), 12, 140–41
Americans for Democratic Action, 62
American Telephone and Telegraph Company, see AT&T
Anderson, Clinton, 88
Antitrust Division, Department of Justice, 30, 41, 173
Antitrust Subcommittee of the Committee on the Judiciary, 27, 52
Apollo moon program, 114, 115, 117, 133
Applications Technology Satellite (ATS) Program, 144–45
Army ADVENT Management Agency, 32
Associated Press, 140
Atlantic City Radio Regulations, 17
Atlas Agena, 110, 113
Atomic Energy Commission, 26

Ball, George W., 49
Bell Telephone Laboratories, 27, 29
Bendix Corporation, 37, 38
Blagonravov, Anatoli A., 125, 128
Boulding, Kenneth, 175
Burdick, 52
Bureau of the Budget, 26

Bureau of Economic Affairs, 48, 50
Busignies, Henri, 37, 39

CETS, see ECSC
COURIER, 20
Cambodia, invasion of, 126
Cannon, 91
Carter, Gilbert, 93, 95, 103, 142
Carter, William, 83
Celler, Emanuel, 53, 65
Charyk, Joseph, 79, 139
Chayes, Abram, 87
Cheprov, I.I., 128–29
Civil-Military relations, 17, 19–21
Clark, Ralph L., 110–11
Cold War, 10, 167
Colino, Richard, 87
Columbia Broadcasting System (CBS), 12, 140
Commerce Committee, 66–67, 87, 147
Committee on Aeronautical and Space Sciences, 64–65, 66, 88
Committee on Peaceful Uses of Outer Space, 124, 125, 127, 128, 129
Committee on Science and Astronautics, 13, 27, 51, 64
Committee on Space Research, 124
Common Market, 129
Commonwealth Conference on Satellite Communications, 92
Communications Policy Board, 40
Communications Satellite Act, 12, 25, 41, 75, 89, 100, 102, 115, 117, 128, 139, 140, 171; passage of, 41–74
Communications Satellite Corporation (Comsat), 3, 4, 41, 69, 70, 71, 73, 80, 114–15, 117, 127, 129, 137–41, 158, 159, 174, 178, 179, 181; relations with Defense Department, 106–13, 118, 119; relations with Executive branch, 81–87, 141–47; relations with NASA, 81–82, 88–89; relations with State Department, 82, 83–86. See also INTELSTAT
*Communications Satellite Report,* 38
Communications satellite systems: alternative, 35–39; demands for, 39–41; early development, 17–46; Interim Arrangements, 75–99; and national security, 105–18; policy and, 21–22, 43–44, 44–46, 171–84. See also ADVENT; Communications Satellite Act; Communications Satellite Corporation; INTELSAT; INTERSPUTNIK

243

Communications Workers of America, 62
Conference of European Postal and Tele-
communications Administration (CEPT),
93, 94
Congressional hearings, Feb.-Apr. 1962,
52-64
Convention of International Liability for
Damage Caused by the Launching of
Objects into Outer Space, 124
Corwin, Edward S., 3
Craven, T.A.M., 18, 38
Czechoslovakia, invasion of, 126

Daddario, 32
Dahl, Robert, 7
D'Arcy, Jean, 166
Declaration of Policy and Purpose in the
National Aeronautics and Space Act,
13-14
Defense Communications Agency (DCA),
111, 117
Defense Satellite Communications System,
113-14, 115, 132
Defense Department, 11, 14, 15, 22, 23, 26,
27, 29, 31, 105-6, 114-15; relations
with Comsat, 106-13, 118, 119; relations
with NASA, 19-21
Deutsch, Karl, 7
Dingman, James E., 28, 37, 39
Docket 11997, 18
Docket 12263, 18
Docket 14024, 28, 30, 33
Docket No. 16058, 117
Docket 11997, 18
Doyle, Stephen E., 131
Dryden-Blagonravov Agreement, 126
Dryden, Hugh L., 61, 90, 125, 128

EARC, 21, 75, 76-78, 86-87, 100, 126, 127,
128, 130
ECHO I, 20, 61, 92
ECHO II, 125, 126
ECSC, 94, 95, 96, 98, 103, 156, 174
EUROSPACE, 92, 165
EUROVISION, 166
Early Bird, 99, 149, 166
Early, Louis B., 87
Ehrlich, Paul, 7
Eisenhower, Dwight D., 10, 13, 22-23, 24,
47
European Conference on Satellite Commun-
ications, see ECSC
European Research Organization (ESRO),
91, 94
European Vehicle Launcher Development
Organization (ELDO), 92, 94, 157
Explorer I, 12

Extraordinary Administrative Radio Conference, see EARC

FCC, 33, 34, 43, 46, 48, 50-51, 60, 65, 67,
68, 75, 79, 118, 139-140, 142-43, 148,
155, 173, 181-82
Farley, Philip J., 31, 40
Federal Communications Act, 14
Federal Communications Commisson, see
FCC
Foreign policy, boundaries of, 180-84
Foreign Relations Committee, 68-69
Frequency allocations, 17-19, 75-78
Fubini, Eugene G., 107, 108, 110, 112, 115
Fulbright, J. William, 86

GE, 12, 25, 27-28, 37, 38, 62
GT&E, 12, 25, 28-29, 36, 138
General Electric Company, see GE
General Telephone and Electric Company,
see GT&E
Geneva Radio Regulations, 78
Glenn, John D., Jr., 125
Glennan, T. Keith, 22
Gore, Albert, 52-68
Government-industry relations, 22-27
Graham, Philip L., 81
Grant, Ulysses, 9
Gruening, 52

H.R. 9696, 52
H.R. 9907, 53
H.R. 10115, 53
H.R. 10138, 52
H.R. 10629, 53
H.R. 10722, 53
H.R. 11040, see Communications Satellite
Act
Harris, Oren, 53-54, 65, 69
Hart, Philip A., 89
Hawaiian Telephone Company (HTC), 12,
62, 138
Holifield, Chet, 111
Hotz, Robert, 123
Hughes Aircraft Company, 12, 37, 62,
63-64, 99
Humphrey, Hubert, 32
Huntington, Samuel P., 80, 171
ICSC, 96, 97-98, 142, 156, 157, 159, 182
INTELSAT, 80-99, 121, 124, 127, 128-30,
131, 132, 134, 135, 171, 172, 174, 176,
178, 179, 181, 182, 183; assessment of,
147-55; interim and definitive arrange-
ments for, 137-70; legal status, 163
IFRB, 78
INTERSPUTNIK, 121, 124, 126, 130-32,
134, 161, 172

INTRAVISION, 166
IT&T, 37, 39, 62, 99, 138, 140, 178
ITT World Com, 12
ITU, 11, 12, 18, 75, 77, 129, 154, 166, 171; Conference, 18-19, 21; Radio Regulations of, 18, 78
*In the Matter of Authorized Entities and Authorized Users Under the Communications Satellite Act of 1962*, 117
"Inquiry into the Administrative and Regulatory Problems Relating to the Authorization of Commercially Operable Space Communications Systems," 24-25
Interdepartmental Drafting Committee, 47-48
Interdepartmental Radio Advisory Committee (IRAC), 18, 75, 79
Interim Arrangements for a Global Commercial Communications Satellite System, 75-104
Interim Communications Satellite Committee, *see* ICSC
Interim Defense Communications Satellite Project (IDCSP), 114
International Business Machines Corporation, 140
International Civil Aviation Organization, 39
International Frequency Registration Board, *see* IFRB
International Geophysic Year, 12, 17, 124
International Radio-Telegraph Union, *see* ITU
International Telecommunications Satellite Consortium, *see* INTELSAT
International Telecommunicatons Union, *see* ITU
International Telegraph Union, *see* ITU
International Telephone and Telegraph Company, *see* IT&T
International Year of the Quiet Sun, 124
Istvan, Edwin J., 84, 85, 140

Johnson, John A., 87
Johnson, Lyndon B., 12-13, 125, 133, 145, 167
Johnson, Roy, 108
Jupiter, 12
Justice Department, 24-25, 26, 29, 30, 31, 43, 46, 48, 50, 51, 59

Karth, 32
Katzenbach, Nicholas, 53, 83
Kefauver, Estes, 32, 52, 53, 54-56, 60, 63, 67
Kennedy, John F., 23-24, 25-26, 28, 42, 43, 45, 47, 52, 81, 107-8, 124, 125, 134

Kennedy, Robert, 51, 52, 55, 56, 59, 63
Kowalski, Frank, 53
Krushchev, Nkita, 124, 125

Lasswell, Harold, 168
Lindblom, Charles E., 7, 174-75, 177
Lockheed Aircraft Corporation, 12, 27, 28, 29, 36
Loevinger, Lee, 30, 41-42, 50
Long, Russell B., 29, 33-34, 41, 55, 68
Low, George M., 126
Lutz, S.G., 63

MACS, 108-9
McConnell, Joseph H., 76, 77-78, 79
McCormack, John W., 13
McElroy, Neil, 20
McGhee, 58, 71
McKinley, William, 9
McLuhan, Marshall, 154
McNamara, Robert, 113
MacQuivey, Donald R., 79
Madison, James, 7
Madrid Conference, 11
Magnuson, Warren, 52, 56, 88
Mann-Elkins Act, 10
Marcuse, Herbert, 7
Medium-altitude communications system, *see* MACS
Meyer, Lewis, 87
Military Operations Subcommittee, 116, 118
Miller, George P., 52
Minow, Newton, 29-30
Moorhead, William S., 116
Morse, Wayne, 32, 52, 68
Murrow, Edward R., 59

NACA, 13, 20
NASA, 11, 14, 15, 17, 26, 28, 29, 30, 36, 37, 43, 46, 50, 60-61, 67, 88, 142, 144, 165, 166, 173; budget, 23-24, 47; relations with Comsat, 82-83, 88-89, 143; relations witn Defense Department, 15, 19-21; relations with FCC, 24
NASC, 11, 25, 26, 27, 50, 51
NASCOM, 116, 117, 147
NCS, 107, 108
National Advisory Committee for Aeronautics, *see* NACA
National Aeronautics and Space Act, 9, 12-15, 17, 19
National Aeronautics and Space Agency, *see* NASA
National Aeronautics and Space Committee, *see* NASC
National Broadcasting Company (NBC), 12, 140

National Association of Manufacturers, 62
National Communications System, see NCS
National Telephone Cooperative Association, 62
Navy, 21
Neuberger, 52
Neumann, John von, 5, 70
Nixon, Richard M., 146, 154, 164, 167
Nuclear Test Ban Treaty, 126

ORBITA, 131, 134, 148
O'Connell, 112–13
Operating Agreement Relating to the International Telecommunications Satellite Organization, 158
Outer Space Treaty, 124, 126

Palma, Samuel de, 155
Pastore, John O., 53, 56–57, 66, 67, 89, 155
Pickering, 133
Philco Corporation, 12, 28, 29, 37, 38
Philco/Space Technology Laboratory team, 107, 108, 113
Pierce, John R., 27
Policy Planning for Space Telecommunications, 20, 40
Policy, U.S., 2–4; adequacy of, 21–22; goals of, 44–46, 171–84; and policy-making, influence of technological changes on, 43–44
Post Roads Act, 9
Preparedness Investigating Subcommittee, 13
President's Science Advisory Committee, 13
Press Wireless, 12

RCA, 10, 27, 36, 62, 99, 138
RCAGC, 12, 140
Radio Administrative Conference, 18
Radio Corporation of America, see RCA
Rae, James R., 108
Ranger, 133
RCA Communications, 25, 41, 62
RELAY, 36, 37, 61
Reiger, Siegfried H., 87
Reynolds Metals, 76
Roback, Herbert, 111, 112, 116
Rome Conference, 95, 96, 97
Rooney, John J., 155
Roosevelt, Theodore, 9
Rostow, Eugene V., 145
Rostow Report, 145–46
Rostow Task Force, 165
Rubel, John H., 61
Ryan, William Fitts, 32, 34, 65

S. 2814, 52, 66

S. 2890, 52
SCORE, 20, 27
SKYNET, 117
SYMPHONIE, 157, 161
SYNCOM, 37, 149
Sarnoff, David, 63
St. Germain, Fernand J., 115–16
Satellites: global commercial system, 75–104; distribution and direct broadcast, 164–68; orbits of, 35–36; passive, 35
Science Advisory Committee, see President's Science Advisory Committee
Scott, Hugh, 34
Seamans, 116
Security, national, 105–20; foreign relations aspects, 114–18
Select Committee on Astronautics and Space Exploration, 13
Senate Committee on Interstate and Foreign Commerce, 27, 33, 34
Senate Space Committee, 21, 39, 65, 66; hearings, 32, 34; report of, 21–22
Six Day War, 126
Smith, David B., 37–38
Soviet Academy of Sciences, 125
Soviet Defense Ministry, 133
Space Committee, see Senate Space Committee
Space Program, development of, 17–46
Space Technology Laboratories, 99
Sputnik, 12, 13, 15, 16, 26, 27, 124
Staggers, Harley O., 155
State Department, 11, 26, 29, 31, 43–44, 48–50, 57–59, 60, 61–62, 79, 141, 155–56
Stevenson, Adlai, 41, 42
Subcommittee on Monopoly of the Select Committee on Small Business, 27, 29–30, 33, 41
Sylvania, 36
Symington, 53, 63, 89

TELSTAR, 36, 37, 61, 67, 92
Teague, Olin E., 52
Telecommunications Coordinating Committee, 18
Thames, William M., 32, 108
Thatcher, 129
Thompson, Llewellyn E., 113
Titan III-C, 110, 113
Transatlantic telephone cable (TAT-1), 46
Truman, David, 7
Truman, Harry S., 10
Twentieth Century Fund, 155

UN, 12, 126, 127
USIA, 59

USSR, 4, 61, 77–78, 79, 106; relations with US on space communications, 121–35
United Press, 140
United States Independent Telephone Association, 62

Vietnam War, 126
Viking, 12

Washburn, Abbott, 155
Webb, James, 30–31, 130
Weisner Committee, 43
Weisner, Jerome B., 23, 83
Weisner Report, 23
Welch, Leo D., 81, 82, 83, 84, 85, 86, 87, 113

Welsh Committee, 47–48, 50–52
West Ford, 20, 105–6, 127
Western Electric, 11, 30, 62, 139
Western Union (WU), 12, 15, 18, 41, 62
Western Union International (WUI), 62
White House Conference on International Cooperation, 165, 167
Whitehead, Clay, 146
World Magnetic Survey, 124
World Meterological Organization, 39
World War I, 9–10
World War II, 10

Yarborough, 52

## About the Author

Jonathan F. Galloway is Associate Professor of Politics at Lake Forest College. A graduate of Swarthmore College, he received his doctorate at Columbia University. He has published articles on multinational corporations and the politics of technological change in such journals as *International Organization, International Studies Quarterly, Journal of Social Philosophy*, and *Politics & Society*. He is currently doing research on the linkages between multinational corporate expansion and military-industrial-technological developments.